KU-567-629

E-literature for Children

Enhancing digital literacy learning

Len Unsworth

Routledge
Taylor & Francis Group

LONDON AND NEW YORK

First published 2006
by Routledge
2 Park Square, Milton Park, Abingdon, Oxon OX14 4RN

Simultaneously published in the USA and Canada
by Routledge
270 Madison Ave, New York, NY 10016

Routledge is an imprint of the Taylor & Francis Group

© 2006 Len Unsworth

Typeset in Sabon by
HWA Text and Data Management, Tunbridge Wells
Printed and bound in Great Britain by
The Cromwell Press, Trowbridge, Wiltshire

All rights reserved. No part of this book may be reprinted or
reproduced or utilised in any form or by any electronic, mechanical,
or other means, now known or hereafter invented, including
photocopying and recording, or in any information storage or
retrieval system, without permission in writing from the publishers.

British Library Cataloguing in Publication Data
A catalogue record for this book is available from the British Library

Library of Congress Cataloging in Publication Data
A catalog record for this book has been requested

ISBN10: 0–415–33329–6 (hbk)
ISBN10: 0–415–33330–X (pbk)

ISBN13: 9–78–0–415–33329–0 (hbk)
ISBN13: 9–78–0–415–33330–6 (pbk)

BLACKBURN COLLEGE
LIBRARY

Acc. No. BB06792
HSC
Class No. 372.64 UNS
Date 15/06/07

Contents

List of figures   vii
List of tables   ix
Preface   xi
Acknowledgements   xv

1 Children's literature and literacy in the electronic age   1

2 Describing how images and text make meanings in
e-literature   13

3 Learning through web contexts of book-based literary
narratives   37

4 Classic and contemporary children's literature in electronic
formats   57

5 Emerging digital narratives and hyperfiction for children
and adolescents   87

6 Electronic game narratives: resources for literacy and
literary development   119

7 Practical programs using e-literature in classroom units
of work   137

References   157
Index   167

Figures

1.1 Describing the articulation of book and computer-based literature   5
2.1 Roger and Ronnie playing tag up and down the aisles   17
2.2 First image in *Wind Song*   21
2.3 Final image in *Wind Song*   21
2.4 The Littlest Knight and his Princess   22
2.5 Strong Framing in *The Littlest Knight*   24
2.6 Framing in *Wolstencroft the Bear*   25
4.1 High angle view: Looking down on the Little Prince   61
4.2 An eye-level view of the geographer   61
4.3 Plate 10 from Wodehouse (1904) *William Tell Told Again*   74
4.4 Margaret Early's image of Gessler ordering Tell to shoot the arrow   75
4.5 Comparing images depicting Tell's second arrow   76
4.6 Dependent clause as Theme   83
5.1 Relating types of e-narratives and compositional features   89
5.2 Wolstencroft mainly associated with 'thinking' and 'feeling' verbs   92
5.3 Reader alignment with Wolstencroft's point of view   93
5.4 Distant view with high vertical angle, oblique horizontal angle constructing remoteness from the reader   96
5.5 Reader alignment with the point of view of the Littlest Knight   97
5.6 Vasalisa site layout   107

Tables

4.1 Animation scenes in chapter 15 of the CD-ROM version of
   The Little Prince   64
4.2 Interactive features of images in the chapter 15 animation of
   The Little Prince   64
4.3 Visual point of view and principal speakers in the scenes of the
   'ephemeral' discussion   66
4.4 Gessler orders Tell to shoot the apple from his son's head   77
4.5 Proportions of different verb types in two versions of
   William Tell   77
4.6 Topical Themes in two versions of an episode in the
   William Tell story   79
4.7 Tell explains his second arrow   80
5.1 Contextual pages linked to *The Inner Circle* online story   101

Preface

Children's literature can bridge the inter-generational digital divide in the English classroom. The digital multimedia world of the world wide web (www) and CD-ROM technology is enhancing and expanding the story worlds of literary narratives accessed primarily through the reading of books, as well as generating exciting new forms of digital narrative such as hyperfiction and electronic game narratives. Some established authors of children's literature such as Jon Scieszka and Paul Jennings and Morris Gleitzman have linked recent book publications to related online games and others like Libby Hathorn have published online literary game narratives like *The Wishing Cupboard*. The burgeoning of children's literature sites on the web reflects not only the popularity of children's books and other forms of literary narratives but also the integral part played by the web in children's experience of such story contexts. But the evidence is that the majority of teachers, even younger, recent graduates, are in need of guidance in seeking to make effective use of the computer facilities that are now widely accessible in their schools and classrooms. At the same time, more and more children routinely use computers outside of school to access a variety of forms of digital narrative on CD-ROM and the web, and more and more they are communicating their experience around story via email, 'blogs' and various forms of electronic forums and chat rooms. There is an opportunity to bring the complementary expertise and experience of children and teachers together in their shared enjoyment of exploring children's literature.

This book shows how teachers can use e-literature in the classroom to enhance and extend the engagement of computer-age children with the enchantment of the possible worlds of literary narratives. It is concerned with the interface of pedagogy, computer technology and children's literature as well as new forms of literary narratives for children on CD-ROM and the web, including hyperfiction and electronic game narratives.

While the orientation is one of practical support for classroom teachers, it is 'research-led' support, reporting the results to date of a range of ongoing studies by the author dealing with the nature of image/text relations and their role in the construction of literary narratives, an evaluation of online resources for developing children's literacy and literary understanding, relationships among conventional book and computer-based versions of ostensibly the same literary narratives, explicating the various types of online digital fiction available for children, and the development of a typology of electronic game activities linked to literary narratives for children. Much of the research methodology, principally web searches and multimodal analyses of digital narratives, involves recent linguistic and visual semiotic description of the meaning-making resources of language and image, which are very accessible tools for literary interpretive work and literacy learning that can be readily mediated to children. Hence their description here is oriented to their use as pedagogic resources. The research outcomes are illustrated with comparative exemplars from the data and summary frameworks to facilitate classroom application as indicated in the following brief chapter outlines.

The first chapter introduces the broad frameworks that will assist teachers in managing effective classroom programs using digital resources for developing literary understanding and literacy learning. These are the **organizational, interpretive** and **pedagogic** frameworks. The *interpretive* framework is described more fully in Chapter 2. This framework addresses the increasingly integrative role of language and images in the construction of literary meanings in electronic and book formats. It describes the verbal and visual grammatical analyses alluded to above in relation to both research methodology and resourcing teachers and students with functional interpretive tools. The *organizational* framework describes the articulation of conventional and computer-based literary narratives for children and adolescents. The main categories of articulation are electronically *augmented*, electronically *re-contextualized*, or electronically *originated* literary texts. Chapter 3 explicates the nature of electronically augmented literary texts describing various ways in which published books, their story worlds, compositional contexts, and readers responses and interactions with them are now extensively mediated via the web. Chapter 4 deals with literary texts in book format which have been re-contexualized in digital form on the web or CD-ROM. Two research outcomes of direct benefit to teachers are reported in this chapter. The first is a documenting of the sources of such re-contextualized literary texts on the web, as well as a listing of some exemplars in CD-ROM format. The second research outcome is an explication of the means by which re-contextualisation processes involving the choices and positioning of images and, to some extent, the editing of text, actually constructs different interpretive possibilities across ostensibly the 'same' story. This kind of explication becomes a key resource for developing active, critical reading among students. Chapters 5 and 6 deal with electronically *originated* literary texts. In the fifth chapter, five types of e-narrative, two types of digital poetry and e-comics are described and illustrated. The types of e-narratives re-defined in terms of their compositional features, inform examples of the types of learning experiences which can be developed around their distinctive digital narrative forms. Chapter 6 distinguishes video games as defined by Gee (2003) from electronic game narratives that are related to literary narratives (which may be either distinct from or integrated with the game). The modest data sample of games on which this research was based has yielded a tentative typology of electronic game narrative activities (summarized in Figure 6.1). This initial account enables teachers to see how various kinds of games can be related in different ways to different dimensions of the stories from which they are derived (or the stories the games themselves actually constitute). Samples of classroom learning tasks based on such relationships are briefly outlined. While Chapters 1 through 6 have been oriented to classroom applications of developing understanding of various aspects of e-literature for children, Chapter 7 draws together these ideas in examples of practical planning of programs of work for children at different levels in the primary/elementary school. The sample programs are presented with a view to encouraging and provoking more innovative explorations by teachers in their own professional practice.

Much more collaborative work is needed in interfacing research and teaching in this area of the English curriculum. The exponential rate of change in the nature of ICTs and their impact on both the nature of literary texts and the contexts in which these texts are experienced, emphasizes the crucial agentive role of both teachers and students in defining and pursing a research agenda of such practical social significance. It is hoped that this book will stimulate critically constructive responses to, and envisioning beyond, what is presented here to maintain and enhance a vibrant engagement of 'net-age' students with past, contemporary and emerging forms of literary narrative.

Acknowledgements

This work would not have been possible without the ongoing remarkable support of my wife, Loraine.

My colleague, Dr Angela Thomas, has sustained and extended my interest in exploring the role of e-literature and online literary resources for children and adolescents through her enthusiasm for, deep understanding of, and incisive scholarship in, the nature of young people's engagement in the digital world.

I should also like to thank the following for permission to use previously published material:

Illustrations from *William Tell* by Margaret Early reproduced with permission by Lothian Books, Melbourne.

Images from *Wolstencroft the Bear*, *Wind Song* and *The Littlest Knight* reproduced with permission of the illustrator, Carol Moore from Children's Storybooks Online (http://www.magickeys.com/books/index.html).

The text from the first two pages of *Wolstencroft the Bear* is reproduced with permission from the author, Ms Karen Lewis. Karen Lewis is the author of several mystery novels, including *A Scarlet Woman* and *A Fatal Affair* published by Treeside Press, *Lingering Doubt* by Felon Books, and *Never Seen Again* by London Circle Publishing – all Detective Neil Slater mysteries. She has written the *Wolstencroft the Bear* and *Elmira the Bear* series for children. She has also written *A Strange Disappearance* for young adults, and her suspense thriller *Hit and Run* won a radio play competition. She makes her home in Vancouver, Canada.

Images from *The Vasalisa Project* reproduced with permission of the author Joellyn Rock (http://www.rockingchair.org/).

Chapter 1

Children's literature and literacy in the electronic age

Introduction

The ways in which children and young people interact with literary texts are being profoundly influenced by the internet and the world wide web (www) as well as other aspects of contemporary Information and Communication Technologies (ICTs). In fact, the impact of ICTs is changing the nature of literary texts and also generating new forms of literary narratives (Hunt, 2000; Locke and Andrews, 2004), including some video game narratives (Gee, 2003; Ledgerwood, 1999; Mackey, 1999; Zancanella *et al.*, 2000). However, it is not the case that the literary interests of the digital multimedia world are replacing books as a presentation format for children's literature (Dresang, 1999; Gee, 2003; Hunt, 2000). Rather, what we see emerging are strongly synergistic complementarities, where the story worlds of books are extended and enhanced by various forms of digital multimedia, and correspondingly, some types of digital narratives frequently have companion publications in book form. In Margaret Mackey's (1994: 15) words 'Cross-media hybrids are every where.' She points out that children come to school already used to making cross-media comparisons and judgements whether the stories are about Thomas the Tank Engine or Hamlet, and that

> To talk about children's literature, in the normal restricted sense of children's novels, poems and picture-books, is to ignore the multi-media expertise of our children.
>
> (Mackey, 1994: 17)

However, 10 years later, although literature for children and young people maintains its significant role in state and national English curriculum documents, such documents are silent about literary narratives in the digital sphere (Locke and Andrews, 2004). There is also relatively little use of ICTs in teaching literary texts in schools, according to national studies in Australia (Durrant and Hargreaves, 1995; Lankshear *et al.*, 2000). On the other hand, online and other digital media resources for working with literature in the classroom are burgeoning, access to appropriate computing facilities in schools in Western countries is becoming routine, and there is an emerging research literature dealing with the interface of ICTs, literature and literacy education (Jewitt, 2002; Locke and Andrews, 2004; Morgan, 2002; Morgan and Andrews, 1999). To bridge the gap between many students' experience of literature in the digital world and their classroom experience, beginning teachers who have more familiarity with ICTs, as well as established teachers, who are less familiar with ICTs but have great expertise and experience in working with literature, need access to *organizational*, *interpretive* and *pedagogic* frameworks that will assist them in managing effective classroom programs using digital resources for developing literary understanding and literacy learning. The purpose of this book is to contribute to the development of these frameworks. This chapter provides an overview of key elements of the frameworks, which are then discussed in more detail in the subsequent chapters, with examples from online narratives, teaching resources and suggestions for teaching/learning activities. The first stage of the overview here deals with an *organizational* framework describing the articulation of conventional and computer-based literary narratives for children and adolescents. The second stage outlines *interpretive* frameworks addressing the increasingly integrative role of language and images in the construction of literary meanings in electronic and book formats. The third stage deals with *pedagogic* frameworks, beginning with the online contexts for developing understanding about different dimensions of literary experience, and then addressing the management of learning activities derived from such contexts in extended programs of classroom work.

Describing the articulation of book and computer-based literary narratives

Here we are concerned with the relationships among literary materials on the web, on CD-ROMs and in books. From time to time in this book mention will be made of movie versions of various literary works and their availability as DVDs, but the focus will be online and CD resources. It is useful to think about the relationships among literary texts and digital media in terms of three main categories. The first refers to electronically *augmented* literary texts, or perhaps electronically augmented experience relating to literary texts. This category is concerned with literature that has been published in book format only, but the books are augmented with online resources that enhance and extend the story world of the book. This kind of augmentation is most frequently provided by the publishers and/or the authors themselves. Sometimes it involves information about the genesis of the story, further details of artefacts or additional information about characters, and sometimes it involves presentation of selections from the story in print or by the author, or someone else, reading a sample chapter or segment, to entice the potential reader to invest in the whole story. The ways in which books are augmented with such online enhancements are discussed in detail in Chapter 3.

The second category of relationship among literary texts and digital media is the electronically *re-contextualized* literary text. In this category, literature that has been published in book form is re-published online or as a CD-ROM. The online re-publication takes a variety of forms. Many works that are now in the public domain because copyright laws no longer apply (usually because it is more than 50 or 70 years since the death of the author) have been transcribed or scanned and located in online digital libraries. The most widely known of these is the Gutenberg Project (http://gutenberg.net/), but there are many other such online libraries, including some specializing in books for children. These resources are detailed in Chapter 4. The scanned books contain the original images, but since copyright is not an issue, some other sites provide the texts of these stories with new images interpolated. These online versions of published books can be accessed free of charge. The second type of online version of published books is usually contemporary stories that are provided by publishers and can be downloaded at a cost. It is also possible, at a modest cost, to download audiofiles for many current titles, including classics like Oscar Wilde's *The Selfish Giant* (Wilde and Gallagher, 1995), as discussed in Chapter 7. Some books are also published as audio only CDs, such as Stephen Fry's reading of the Harry Potter books, published by BBC Audio Books in the United Kingdom. But most CD-ROM versions of literary texts include images and text, which vary to a greater or lesser extent from those in the book versions. In some cases the images are static, simply transposed from page to screen. This is the case with *The Paper Bag Princess* (Munsch, 1994) for example. In other cases the original images from the book appear as animations on the CD as in *The Polar Express* (Van Allsburg, 1997). In this CD the animations activate automatically, but in others like *The Little Prince* (de Saint-Exupery, 2000b), the animations are controlled by the mouse 'clicks' of the viewer. In some cases novels for mature readers such as *Of Mice and Men* (Steinbeck, 1937; Steinbeck Series, 1996) have been re-presented as CD-ROM versions including images throughout. Literature *re-contextualized* as CD-ROM presentations is discussed in Chapter 4.

The third category relating literary narratives to digital format is the digitally *originated* literary text. These are stories that have been published in digital format only – on the web or CD-ROM. Relatively few such stories appear on CD-ROM. Some notable examples (James, 1999) such as *Lulu's Enchanted Book* (Victor-Pujebet, n.d.) and *Payuta and the Ice God* (Ubisoft, n.d.) are discussed in Chapter 5. The great variety of literary narratives for children and adolescents published on the web can be categorized as follows:

- *e-stories for early readers* – these are texts which utilize audio combined with hyperlinks to support young children in learning to decode the printed text by providing models of oral reading of stories and frequently of the pronunciation of individual words;
- *linear e-narratives* – these are essentially the same kinds of story presentations which are found in books, frequently illustrated, but presented on a computer screen;
- *e-narratives and interactive story contexts* – the presentation of these stories is very similar to that of linear e-narratives, however the story context is often elaborated by access to separate information about characters, story setting in the form of maps, and links to factual information and/or other stories. In some examples it is possible to access this kind of contextual information while reading the story;
- *hypertext narratives* – although frequently making use of a range of different types of hyperlinks, these stories are distinguished by their focus on text, to the almost entire exclusion of images;
- *hypermedia narratives* – these stories use a range of hyperlinks involving text and images, often in combination.

To this list must be added some types of video games, defined in Chapter 6 as electronic game narratives. The development of new forms of literary narrative in the context of electronic games has been a focus of a recent study (Ledgerwood, 1999; Mackey, 1999; Murray, 1998; Zancanella *et al.*, 2000). Examples of these forms of e-fiction as well as e-poetry and e-comics are discussed in Chapter 5, and electronic game narratives are discussed in Chapter 6.

All three of the above categories relating literature to the resources of the web and CD-ROM technology vary from monomodal (print only) to multimodal presentation, involving print, images and sound. The digitally *re-contextualized* and digitally *originated* e-fiction also vary along the continua of linear to hyperlinked and from conventional story structure to innovative game narratives. The *organisational* framework describing the articulation of book and computer-based literary texts for children and adolescents is summarized in Figure 1.1.

Interpreting the joint role of images and text in constructing literary narrative

Over the last decade images have become increasingly prominent in many different types of texts in paper and electronic media. Recent publications of popular fiction and new editions of classic literature are now frequently richly illustrated. This can be seen in novels such as Terry Pratchett's Discworld Fable *The Last Hero* illustrated by Paul Kidby (2001), and the edition of Tolkien's *Lord of the Rings* illustrated by Alan Lee (2002), as well as in illustrated novels for young readers such as Isobelle Carmody's *Dreamwalker*,

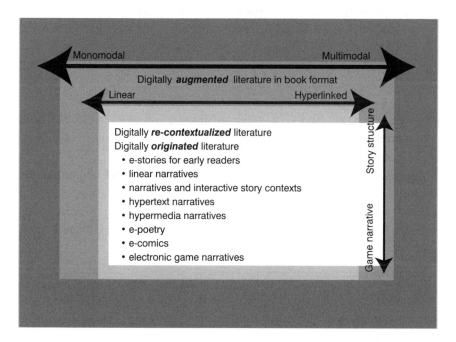

Figure 1.1 Describing the articulation of book and computer-based literature

illustrated by Steve Woolman (2001). The kinds of images and their contribution to overall meaning vary with the type of narrative. However, overwhelmingly, both the information in images and their effects on readers are far from redundant or peripheral embellishments to the print. Because images are used increasingly, and in a complementary role to the verbal text, it is now inadequate to consider reading simply as processing print. The need to redefine literacy and literacy pedagogy in the light of the increasing influence of images is widely advocated in the international literature (Andrews, 2004a, 2004b; Cope and Kalantzis, 2000; Goodman and Graddol, 1996; Lemke, 1998a, 1998b; Rassool, 1999), drawing attention to 'the blurring of relations between verbal and visual media of textuality' (Richards, 2001). Writing about *Books for Youth in a Digital Age*, Dresang noted that:

> In the graphically oriented, digital, multimedia world, the distinction between pictures and words has become less and less certain.
>
> (Dresang, 1999: 21)

and that

> In order to understand the role of print in the digital age, it is essential to have a solid grasp of the growing integrative relationship of print and graphics.
>
> (Dresang, 1999: 22)

There is also a strong consensus that the knowledge readers need to have about how images and text make meanings, both independently and interactively, requires a metalanguage, or a grammar, for describing these meaning-making resources.

In the book *Tellers, Tales and Texts* (Hodges *et al.*, 2000) Bearne noted that:

> Once readers develop a metalanguage through which to talk about texts they are in a position to say – and think – even more.
>
> (Bearne, 2000: 148)

And earlier in his classic study, *Words About Pictures: The Narrative Art of Children's Picture Books*, Perry Nodelman noted that the interpretation of the narrative role of images in children's books would be enhanced by the possibility of a system underlying visual communication that is something like a grammar – something like the system of relationships and contexts that makes verbal communication possible.

(Nodelman, 1988: ix)

The development of systemic functional linguistics (SFL) (Halliday and Matthiessen, 2001; Martin, 1992; Martin and Rose, 2003; Matthiessen, 1995) and its application to work with literature for children (Austin, 1993; Hasan, 1985; Knowles and Malmkjaer, 1996; Stephens, 1994; Williams, 2000), as well as the extrapolation from SFL of a grammar of visual design for reading images by Kress and van Leeuwen (1996) and its application to work with chldren's literature (Lewis, 2001; Stephens, 2000; Unsworth, 2001; Unsworth and Wheeler, 2002; Williams, 1998) has brought Nodelman's earlier wishes to reality. The interpretive frameworks offered by this work, and their use in understanding the role of images and text in constructing meanings in literary narrative, will be introduced in Chapter 2.

Towards a pedagogic framework for e-literature and classroom literacy learning

The interpretive tools provided by functional descriptions of verbal and visual grammar introduced in Chapter 2 enable teachers and students to read literary texts grammatically, so that they are able to read the 'constructedness' of the texts, simultaneously focusing on the 'what' of the story and the 'how' of its verbal and visual construction. Throughout this book, suggestions for learning experiences designed to support young readers in developing this interpretive, grammatical reading will be introduced in the context of the foci of the subsequent chapters – classic and contemporary stories online, emerging literary hypertext for children and electronic game narratives. This perspective on developing children's literary understanding and concomitant literacy development is a particular innovative feature of the research reported in this book and does not currently find explicit expression in the online resources for using e-literature in the English curriculum. Nevertheless, there are richly inspiring online resources for extending children's literary experience, and the approach in this book is to co-opt such resources for infusion with the above perspective forming a basis for enhancing children's experience of e-literature in school contexts. This section of the chapter addresses the nature of a pedagogic framework for using e-literature in the classroom in two stages. The first stage will briefly indicate the range of online contexts for developing understanding about different dimensions of literary experience. These will be explicated in more detail in Chapter 3 and their application in relation to the specific foci of the subsequent chapters will be addressed in those chapters in turn. The second stage of the pedagogic framework deals with principles for the design and implementation of coherent classroom programs of work over extended periods of about two weeks, incorporating the types of learning experiences introduced previously, in focused studies of literary texts involving both close and broad reading experiences. These principles underpin the examples of classroom programs described in detail in Chapter 7.

Online contexts for developing different dimensions of engagement with e-literature

Our engagement with literary texts online may involve one or more of a range of dimensions, each with varying degrees of involvement. Five such dimensions are briefly described here:

*   *Composition/story genesis.* This includes information about actual events, places, artefacts etc., which the author drew on in composing the story. It could also include manuscript data about earlier drafts and episodes/ events/ characters that were excluded or changed, as well as additional information provided by the author to elaborate aspects of the story world constructed in the narrative.
*   *Invitation/enticement to read.* The web provides 'teaser' sample chapters/ segments of stories, often available with audio and sometimes with the author as reader, as well as online reviews and reactions from readers, and more recently online story-derived games designed to arouse reader interest in the narrative.
*   *Appreciation/celebration.* There are many examples of 'fan' sites on the web where individuals or groups of readers manage a site that celebrates a particular author and his/her work. These often contain biographical information, testimonials to the impact of books, favourite quotations, images of covers of different editions and a range of other features, which are listed below in other dimensions of literary engagement.
*   *Interpretation/response.* Two main types of online resources offer opportunities for interpretive responses to the narrative. One type is the fairly traditional lesson plans and learning tasks for teachers to download,

although some of these include online learning experiences that make more use of the affordances of the online digital environment. The second type is the opportunity for readers to participate in online discussions about the books they have been reading via chat rooms and forums.

• *Adjunct composition/creation*. This kind of engagement is frequently evidenced on 'fan' sites where contributors to the site write stories in the style of particular narratives, sometimes additional episodes, sometimes parallel or related stories, often involving the same characters as the original. Some fan sites conduct competitions involving this kind of writing, with strict rules relating the new fiction to parameters of the source story. Other contributions include the creation of images, games and puzzles based on the stories. Another kind of creative composition activity adjunct to the source story is the co-creation of multimodal story episodes in virtual worlds known as palaces. Story palaces, described in Chapter 3, involve participants adopting character roles and representing these characters visually on screen using 'avatars' as well as verbally by the input of dialogue, so that they 'act out' stories in this multimodal virtual world.

Principles for the design and implementation of coherent classroom programs of work

The pedagogic framework involves the strategic use of student-centred, discovery learning as well as teacher directed, overt teaching and intermediate, guided investigations of various kinds. Managing classroom learning also includes designing learning experiences based on collaborative small group activities, individual independent work and common whole class tasks. The teacher at times will be a facilitator and guide or a co-researcher, but at other times will be an authoritative (but not authoritarian) leader and direct instructor. Initial work on a topic, for example, may involve sharing of informal knowledge, observations, and opportunities and suggestions for extending understanding. This may be highly student-centred and exploratory but as the teacher begins to bridge towards negotiating more systematic knowledge, the pedagogic orientation shifts to more guided investigation and direct instruction. On the basis of students' greater familiarity with systematic knowledge of the topic the teacher then moves to emphasize more critical framing to provoke critical questioning by students and a shift towards transformative knowledge. This kind of work may entail more collaborative group work and independent research and may also a shift back to more student-centred, student-initiated learning. As the classroom work progresses through these phases, teaching is differentiated to optimize the engagement of all students in essentially the same learning tasks. This means sophisticated planning and preparation. It might include providing scaffolded learning guides for some students. It could also involve grouping students with high support needs together to 'prime' their understanding of subsequent tasks through direct teaching while more proficient learners operate independently. Then regrouping students heterogeneously so that highly proficient students and high support students are able to work productively together on collaborative tasks. The following principles of dynamic, functional organisation of whole class, small group and individual learning underpin the implementation of the programs described in Chapter 7.

Composition of groups

- There are no permanently set groups.
- Different classroom groupings of students are formed from time to time on varying functional criteria, which may sometimes include proficiency in managing particular learning tasks.
- Groups are periodically created, modified or disbanded to reconcile changing student needs and task demands.
- There are times when there is only one group – consisting of the entire class.
- Group size will vary from two or three to nine or ten or more, depending on the group's purpose.

Management

- Students should encounter some choices relating to how, when and what work is to be done so that they can exercise some responsibility in organizing their own learning.
- Student commitment is enhanced if they know how the group work is related to the overall class program.
- There should be clear, context-sensitive strategies for monitoring and recording students' participation and progress in learning activities.
- There should be a principled distribution of teacher-interactive, teacher-supervized and teacher-independent tasks among group work activities.

Scaffolding learning tasks

- Learning experiences should be based on functional texts in genuine communicative contexts rather than fragmented exercises.
- Task structure, level of demand and forms of scaffolding should be differentiated to accommodate the range of experience and proficiency among students in the class, ensuring that all students are supported in improving their learning.
- Clear models and demonstrations of work required should be provided.
- Directions for completing tasks should be clear and able to be referred to if forgotten.
- Group tasks should lead to some form of corporate production, display or exchange involving the whole class and sometimes interaction with other groups within and/or outside the school.

Because the 'group-then-regroup' strategy (Unsworth, 1993) alluded to above is so fundamental to the practical success of the approach to classroom work adopted in this book, its key elements are emphasized here. The objective is to provide support for less proficient or less experienced learners so that they can participate with more proficient learners in learning to learn from a range of online and offline resources which they would have difficulty dealing with independently. This involves a two-stage learning activity where the first stage is preparatory to the second. In the first stage the students are grouped homogenously according to relative proficiency. The less proficient students are given supportive versions of the learning tasks containing additional guidance and/or are supported by explicit instruction from the teacher in the first phase of the task. At the same time the more proficient learners complete the task without such scaffolding. The students are then re-grouped and asked to undertake a follow-up or review task that relies on the understandings developed in phase one. Because the less proficient students have been supported in the first phase they are better able to participate collaboratively with their more proficient peers in the second follow-up or review phase.

Resourcing scholar-teachers as researching practitioners

Literature for children and young people, and e-literature in particular, is a rapidly evolving social phenomenon, now intimately related to advances in

ICTs. A growing number of educators are now advocating the need for curriculum design and classroom teaching to be responsive to these changes and, in so doing, to acknowledge the relevant experience and expertise of children, which many adult educators do not possess (Alvermann, 2004; Andrews, 2004a; Chandler-Olcott and Mahar, 2003; Gee, 2003; Lankshear and Knobel, 2003; Lankshear *et al.*, 2000; Sefton-Green, 2001; Sefton-Green and Buckingham, 1998). However, it is also obviously the case that teachers are in a position to mediate areas of new knowledge and understanding that are not so readily accessible to children. An example relevant to this work is the theoretically articulated functional grammars of language and image that facilitate explicit discussion about relationships among narrative form and the interpretive possibilities constructed by the multimodal texts. The need to bring together the differential expertise of teachers and students, as well as writers, multimedia composers and researchers, underlines the crucial role of teachers in classroom-based research in addressing the theoretical and practical agendas concerning the role of e-literature in a futures oriented curriculum (Locke and Andrews, 2004). This book has been designed to assist teachers in considering this kind of role orientation. It provides, through the *organizational* framework, a basis for teachers to take a more informed position in negotiating with students the nature of learning to develop literary understanding in the online world of digital multimedia. Through the *interpretive* framework, it provides access to some key aspects of facilitating knowledge that teachers need in order to mediate metasemiotic understanding to students, as a practical tool for interpreting multimodal literary narratives. Also through the *pedagogic* framework it takes account of the pragmatic prominence for teachers of managing the day-to-day classroom practicalities in providing learning experiences to meet the differential needs of the range of students they meet in any one teaching context. Like any book, this volume can be browsed, and readers may order their own selection of the chapters, however the sequencing of chapters is intended to assist those for whom e-literature in the classroom is a relatively unfamiliar experience, and even for those more familiar with the topic, it is suggested that at least the first two chapters are read consecutively and prior to the subsequent chapters.

Chapter 2

Describing how images and text make meanings in e-literature

Introduction

Many children and teenagers are actively and independently engaged in exploring e-literature. This is evidenced in the subsequent chapters which indicate not only the variety of responses to stories that young people post on the web, but also the fact that many websites devoted to favourite literary authors are actually constructed and managed by young people. The literary activities of these online communities are not necessarily related to students' experience of literature in school contexts (although other sites are clearly oriented to school sponsored participation). What is also clear in the story reviews and character analyses provided on many such sites, as well as the popularity of sites providing multimedia 'back-stories' and para texts, is the interest that young readers have in the 'constructedness' of the stories. It should not be surprising therefore that, far from lessening children's enjoyment of literature, analysing the means by which images and language make meanings helps them feel they are getting closer to the texts and what it is they enjoy about them (Misson, 1998; G. Williams, 2000; M. Williams, 2002). To help children learn how to analyse the ways language and images make meaning, there is growing advocacy of the role of metalanguage – grammatical descriptions of visual and verbal meaning-making resources (Bearne, 2000; Doonan, 1993; Jewitt, 2002; Nodelman, 1988; Quinn, 2004; Russell, 2000; Schleppegrell et al., 2004). But what kind of visual and verbal grammar is most appropriate for analysing multi-modal literary texts for children? This chapter will, first, briefly outline the requirements of such a grammar and indicate how these might be addressed by systemic functional linguistic (SFL) accounts of a functional grammar of language (M. A. K. Halliday and Matthiessen, 2001) and the related 'grammar of visual design' (Kress and van Leeuwen, 1996), which was derived from basic SFL principles.

The practical application and pedagogic usefulness of the visual grammar and the functional verbal grammar will then be illustrated using images and text examples from e-literature for children.

A metalanguage of multiliteracies

In order to develop critical multiliteracies practices (Cope and Kalantzis, 2000; Unsworth, 2001), students need to understand how the resources of language and image can be deployed independently and interactively to construct different kinds of meanings. This means developing knowledge *about* linguistic and visual meaning-making systems and the capacity to use these systems to analyse texts. This entails metalanguage – language for describing language, images and meaning-making intermodal interactions. Metalanguage, in the form of a range of different types of grammar and descriptions of text structure, is not new. Various forms of metalanguage describing technical aspects of images and their production are well known. But what is needed is a metalanguage that describes the 'grammar', or structural elements and their relationships, of images and language in terms of the functions or meaning-making roles of such elements and relationships. This means a metalanguage in which meaning-making in social contexts is fundamental to its technical description of language and image.

The importance of a metalanguage for developing multiliteracies is very widely acknowledged, and there seems to be growing consensus about the kind of metalanguage that is needed. A group of 10 academics, identifying themselves as The New London Group and including members from the UK, the USA and Australia, addressed this issue in their proposal for a pedagogy of multiliteracies (New London Group, 1996, 2000). They emphasized that this metalanguage is needed to support a sophisticated critical analysis of language and other semiotic systems, yet it should not make unrealistic demands on teachers and students. Primarily, however, this metalanguage needed to derive from a theoretical account that links the meaning-making elements and structures of semiotic systems like language and image to their use in social contexts.

> ... the primary purpose of the metalanguage should be to identify and explain differences between texts, and relate these to the contexts of culture and situation in which they seem to work.
>
> (New London Group, 2000: 24)

This aligns with a fundamental premise of SFL, which posits the complete interconnectedness of the linguistic and the social (Halliday, 1973, 1978, 1994; Halliday and Hasan, 1985; Hasan, 1995; Martin, 1991, 1992).

Extrapolating from SFL descriptions of language, researchers have developed a corresponding functional account of 'visual grammar' (Kress and van Leeuwen, 1990, 1996; Lemke, 1998b; O'Toole, 1994). This work recognizes that images, like language, realize not only representations of *material reality* but also the interpersonal interaction of *social reality* (such as relations between viewers and what is viewed). The work also recognizes that images cohere into textual compositions in different ways and so realize *semiotic reality*. More technically, functional semiotic accounts of images adopt from SFL the metafunctional organisation of meaning-making resources:

- *Representational/ideational* structures verbally and visually construct the nature of events, the objects and participants involved, and the circumstances in which they occur.
- *Interactive/interpersonal* verbal and visual resources construct the nature of relationships among speakers/listeners, writers/readers and visuals/viewers.
- *Compositional/textual* meanings are concerned with the distribution of the information value or relative emphasis among elements of the text and image.

The New London Group indicated that what is needed to support a pedagogy of multiliteracies is

> ... an educationally accessible functional grammar; that is, a metalanguage that describes meaning in various realms. These include the textual and the visual, as well as the multimodal relations between different meaning-making processes that are now so critical in media texts and the texts of electronic multimedia.
>
> (New London Group, 2000: 24)

We do not yet have such an integrative, intermodal grammar. However, recent work in educational contexts with systemic functional grammar and the related 'grammar of visual design' has shown how students can be taught to be alert to the multimodal 'constructedness' of texts (Astorga, 1999; Callow and Zammit, 2002; Humphrey, 1996; D. Lewis, 2001; Stephens, 2000; Styles and Arizpe, 2001; Unsworth, 1999, 2001, 2003; van Leeuwen and Humphrey, 1996; Veel, 1998; G. Williams, 1998, 2000). The following sections will outline basic concepts in these accounts of visual and verbal grammar, indicating their application to the images and language of e-literature.

Rethinking reading narrative images: using a grammar of visual design

A key pedagogic advantage in using the systemic functional grammar of language and the grammar of visual design is the common theoretical basis that three kinds of meanings are always being made simultaneously in language and in images. This means that the analyses of language and images can be compared in terms of their contribution to the construction of the different kinds of meanings. This section will illustrate the ways in which this metafunctional framework can describe visual meaning making in e-literature. Although the three metafunctions are realized simultaneously, we will initially discuss each separately. First, we will consider aspects of *representational* structures, which visually construct the nature of events, the objects and participants involved and the circumstances in which they occur. Second, we will examine the construction of *interactive* meanings in images, which include the interpersonal relationship between the viewer and the represented participants. Then we will investigate how aspects of layout construct *compositional* meanings, which are concerned with the distribution of the information value or relative emphasis among elements of the image.

Representational meanings

According to Kress and van Leeuwen (1996), images construct representations of reality that are either 'narrative' or 'conceptual'. Narrative images can depict participants (human or non-human) participating in actional, reactional, verbal events or mental events (the latter by means of 'thought clouds'). Sometimes several of these processes occur in the same image. Representational images are characterized by the presence of 'action lines' or 'vectors' and/or speech balloons and thought clouds. Action lines can be seen in Figure 2.1 from the e-story *Wolstencroft the Bear* (K. Lewis and Moore, 2003). These are mainly indicated in the angle of the outstretched legs and the 'swept back' ears, as well as the posture of the arms. The running processes in this image are actional because the participants are not acting upon anyone

Figure 2.1 Roger and Ronnie playing tag up and down the aisles

or anything else. However, such processes can be 'transactional' if the participant is acting on someone or something else. For example, in image (a) in Figure 4.5 (see p. 76) from the story of *William Tell* (Early, 1991) there are a number of transactional actions, such as the soldiers arresting Tell, as indicated by the angle of their arms apparently holding his arms behind his back and the other soldier pointing a sword at Tell.

Reactional processes are when the participant's eyeline is directed to another participant. In Figure 5.3 fom *Wolstencroft ...* (K. Lewis and Moore, 2003) the bears are reacting to their friends being taken away in shopping bags by the young couple. The bears' eyelines are both directed to these participants. Their friends in the shopping bags are also reacting to the bears left behind since the eyeline of the one in the left shopping bag is directed to Wolstencroft and the eyeline from the other shopping bag is directed to the other bear, Rita. These reactional processes are transactional because both the reacting participant and the phenomenon appear in the same image. Reactional processes can also be non-transactional when the phenomenon the participant is reacting to is not shown within the image. This is the case in the third image in *Where the Wild Things Are* (Sendak, 1962) where Max is looking with annoyance to the left of the image, but we are not shown at what. Such images are an important narrative technique in creating the 'gaps' to be filled by the active reader. In this case we are provoked to infer what Max is reacting to.

Verbal processes are included in images as the spoken language printed inside 'speech bubbles'. This is well-known from cartoons, including online cartoons (http://www.onlinecomics.net/pages/index.php), but speech bubbles also occur in other types of online literary narrative, such as the traditional European tales at *A Europe of Tales* website (http://www.europeoftales.net/site/en/index.html). Thinking processes are sometimes included in images within 'thought clouds' – see for example Scott McCloud's *My Obsession with Chess* (http://www. scottmccloud.com/comics/chess/chess.html).

Conceptual images depict classifications or part-whole relations or symbolic relations. Classificatory or part-whole images are not found very frequently in literary narratives and will not be discussed here, although it may be useful to note some examples that do occur. One well-known instance is in Anthony Browne's *Piggybook* (1986), where the classificatory images of 'Mum' show the different kinds of housework she does during the day. A further example is in the *William Tell* picture book by Margaret Early (1991), where the initial image, divided into quadrants, shows various types of terrorism inflicted by Gessler's soldiers on the people in Tell's village. Elsewhere (Unsworth, 2001) I have discussed symbolic images in picture books such as the use of symbolic attributes as in the image in Anthony Browne's *Zoo*, which shows the cloud formations resembling horns on the side of Dad's head. Sometimes the entire image is symbolic as in the case of the penultimate image of the empty chair in John Burningham's *Grandpa* (1984). A further interesting use of a symbolic attribute occurs in the recent online story *Tiger Son* (Ng, 2004). This story, based on an ancient Chinese tale, is set in a time when it was the practice to hunt tigers in this area of China to obtain various products such as skins, meat and bones. An elderly widow's son was killed during one hunt and, to provide compensation, the other hunters gave her a cub, which she could rear and then kill for the profits. She did not kill the cub but freed it when the tiger grew to adulthood. This tiger nevertheless protected the old woman and guarded her grave for a long time after her death. It is only in the final image of the tiger guarding the grave that it appears with a tag hanging around its neck, symbolizing its alliance with at least one human.

Interactive meanings

We will outline three main aspects of the account of interactive meaning in images provided by Kress and van Leeuwen (1996). The first of these is the kind of contact between the viewer and the represented participants in the image – whether the viewer interpersonally interacts with or observes the represented participants. The second aspect is the social distance – whether the image is located along a continuum characterized by a close-up, medium or long shot. The third aspect is the interpersonal attitude that is constructed by the vertical angle (high, medium or low shot) and the horizontal angle (whether the representation is from an oblique angle or whether the viewer seems to be parallel or 'front on' to the image).

One kind of contact between a viewer and the represented participants is referred to by Kress and van Leeuwen as a 'demand'. This is where the gaze of one or more of the represented participants is directly towards the viewer. The second image in the *Wolstencroft...* story (K. Lewis and Moore, 2003) portrays the main character via a 'demand' image as shown in Figure 5.2. This has the effect of optimizing the engagement of the reader with Wolstencroft because he is the only character looking directly at you as the reader. All of the other bears on the shelf are looking elsewhere. Images in which there is no direct gaze toward the viewer from any of the represented participants are referred to by Kress and van Leeuwen as 'offers'. These are images that invite the viewer to contemplate what is occurring rather than interact directly with represented participants in a kind of pseudo interpersonal engagement. Examples of offers can be seen in Figures 5.3 and 5.4.

The concept of social distance in images refers to the continuum of close-up, medium and distant views of represented participants. Close-up views portray the head and maybe the shoulders of a participant only and hence simulate quite an intimate social space between the viewer and the participant in the image. This can be seen in Figure 5.3 where the close-up view of the side of Wolstencroft's head positions the viewer as being quite closely alongside him. A social or medium view portrays the upper half of the participant's body. More distant views include the whole of the participants' bodies as shown in Figures 4.3 and 4.4. More remote images represent the whole bodies of the participants as being relatively very small in relation to the remainder of the image setting and hence position the viewer as being a long way away. This kind of remote view can be seen in Figure 5.4.

The vertical angle of the image accords relative power to the viewer or the represented participant(s). Low angle views show the represented participants looking down on the viewer and hence the represented participant is accorded more power the lower this vertical angle. Figure 4.3 from *William Tell Told Again* (Wodehouse, 1904) shows Gessler mounted on his horse issuing his command to Tell about shooting the apple from his son's head. As readers, we appear to be looking up at Gessler on his horse, so like Tell, who is on foot in the image, we are positioned as less powerful and somewhat vulnerable relative to Gessler.

When the viewer is 'front on', or the horizontal angle of the image is parallel, the effect is to maximize the viewer's involvement or identification with the world of the represented participants. This is the predominant choice of horizontal angle for the images in he *Wolstencroft* ... story. In Figure 5.2 our frontal plane is parallel to that of Wolstencroft, and in Figure 5.3 it appears that we are standing beside him and, like him, our frontal plane is parallel to that of the young couple and the bear's friends who are being carried away in the shopping bags. On the other hand, the first and last images in *Wind Song* (Moore, 2001) shown in Figures 2.2 and 2.3 respectively show the represented participants at an oblique angle. We do not look at them 'front on' but from an oblique angle, so it is as if we are not as involved or included in their world.

One further aspect of interactive meaning described by Kress and van Leeuwen (1996) is modality. This refers to the extent to which an image depicts the realism of the aspects of the world it represents and is determined by the benchmark of the high quality colour photograph. People judge an image to be 'naturalistic' if it approximates this level of representation. Colour is a major influence on naturalistic modality. Naturalistic images have high colour saturation rather than black and white. Their colours are diversified rather than monochrome, and they are modulated, using many shades of the various colours. Modality also varies along a scale from maximum delineation of detail features of participants to the schematization of detail. In highly schematic images a head may be represented by a circle, the eyes by two dots, and the mouth by a curved line. The image of Vasalisa in Figure 5.6 is fairly schematic and of relatively low modality compared with the second last image from *The Littlest Knight* (Moore, 1994) shown as Figure 2.4. The images of the Littlest Knight and his Princess are realistic, if not naturalistic. On the other hand, the online story, *The Wire* by John Marsden (http://www.panmacmillan.com.au/johnmarsden/jmwire1a.htm) is richly illustrated

Figure 2.2 First image in *Wind Song*

Figure 2.3 Final image in *Wind Song*

Figure 2.4 The Littlest Knight and his Princess

with naturalistic photographs. It might be argued that their modality is lowered because they are black and white, but they are also images from the frontline of an invasion and this perhaps marks them as historically 'authentic' and hence increasing their modality from that perspective.

Compositional meanings

Compositional meanings principally concern the ways in which features of 'layout' of the page or screen function as resources for organizing the representational and interactive meanings into a coherent composition that is meaningful as a whole. These include framing and the use of borders of various kinds around the screen as a whole and around images and/or text blocks within screens. They also include resources for indicating the visual importance or salience of various elements and the impact of the placement of elements to the left or right and top or bottom of the screen.

The use of very distinct borders constructs 'strong' framing so that elements marked off by such framing are maximally 'disconnected' from each other. In other words the framing influences the 'reading' of these elements as distinct and relatively less connected to the remainder of the layout. This can be seen in *The Littlest Knight* (Moore, 1994). Figure 2.5 shows page eight of the online story. Here the image has a very distinct line border, which marks it off from the text. This is emphasized by the different background colour for the text and the image. In contrast, 'weak framing' implies maximal integration of elements of layout. This can be seen in the *Wolstencroft...* story (K. Lewis and Moore, 2003). On the first page, shown in Figure 2.6, there is no distinct border around the truncated image of the bear. Although the difference in background colour and location of the image does mark it off from the text, the image is more connected to the text than is the case in *The Littlest Knight*. In fact, in *Wolstencroft* this 'fuzziness' of the image frame is always the case when the focus of the image is Wolstencroft as seen in Figure 5.2 showing Wolstencroft on the shelf. However, when the focus is on the other characters in the story the border becomes more distinct. This can be seen in Figure 2.1 showing Ronny and Roger, the rabbits, running up and down the aisles and also in Figure 5.3 showing Roger and Ronnie being carried away in the shopping bags. Towards the end of the story, in the image showing Rita reading to Wolstencroft on page five, and in the image of Wolstencroft being carried by his new owner, there are no borders at all and the background to the images is the same as that of the text, so there is much greater connection between the images and the text. It seems then that framing is used to emphasize Wolstencroft as the focalizing character by connecting his images maximally with the text. Similar framing effects can be observed in other online stories. In the Australian Aboriginal folktales on the *Bunyips* site (http://www.nla.gov.au/exhibitions/bunyips/) for example, the images are strongly framed and maximally disconnected from the text. On the other hand, in the story of *Banpf* (Left Handed Creations, 1994–2004) the images are, for the most part, not framed and are similar to the latter two images referred to in the *Wolstencroft* story.

Establishing the salience of particular elements can be achieved in a variety of ways. In Figure 2.6 the importance of the name tag for Wolstencroft is achieved partly by the truncating of the rest of the image of the bear and hence giving visual prominence to the tag. The tag therefore takes up more of the image space and it is also written in the distinctive script. In Chapter 4 we will discuss an episode in the CD-ROM story of *The Little Prince* (de Saint-Exupery, 2000b) where the Little Prince has a discussion with the geographer about the geographer's job and also, more philosophically, about what 'ephemeral' means. When the discussion is about the geographer's role

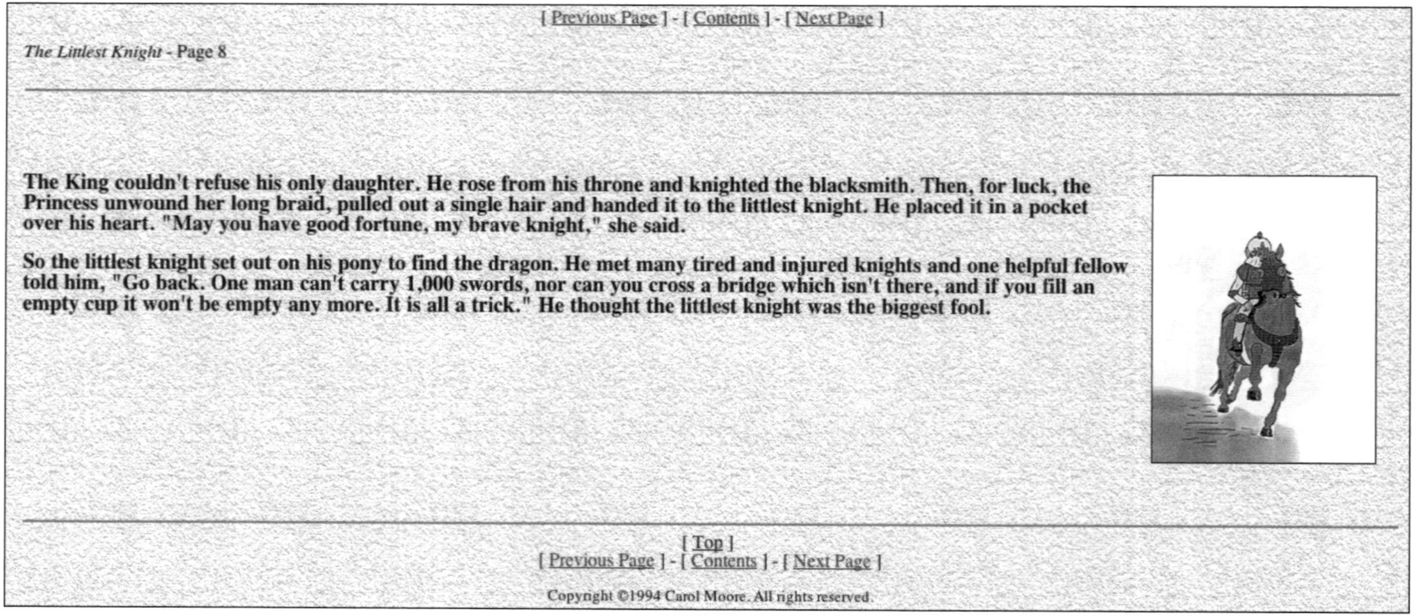

[Previous Page] - [Contents] - [Next Page]

The Littlest Knight - Page 8

The King couldn't refuse his only daughter. He rose from his throne and knighted the blacksmith. Then, for luck, the Princess unwound her long braid, pulled out a single hair and handed it to the littlest knight. He placed it in a pocket over his heart. "May you have good fortune, my brave knight," she said.

So the littlest knight set out on his pony to find the dragon. He met many tired and injured knights and one helpful fellow told him, "Go back. One man can't carry 1,000 swords, nor can you cross a bridge which isn't there, and if you fill an empty cup it won't be empty any more. It is all a trick." He thought the littlest knight was the biggest fool.

[Top]
[Previous Page] - [Contents] - [Next Page]

Copyright ©1994 Carol Moore. All rights reserved.

Figure 2.5 Strong Framing in *The Littlest Knight*

N ot long ago and not far away there was a beautiful, big teddy bear who sat on a shelf in a drug store waiting for someone to buy him and give him a home.

His name was Wolstencroft. And he was no ordinary bear.

His fur was a lovely shade of light grey, and he had honey colored ears, nose and feet. His eyes were warm and kind and he had a wonderfully wise look on his face.

Wolstencroft looked very smart in a brown plaid waistcoat with a gold satin bow tie at his neck.

Attached to the tie was a tag with his name written in bold, black letters: **Wolstencroft.**

Figure 2.6 Framing in *Wolstencroft the Bear*

the images of the Little Prince and the geographer occupy a small space on either the left of right of the screen. However, when the discussion is about the meaning of 'ephemeral' the image of the Little Prince or the geographer occupy the full screen. More prominence is given to the characters in this latter discussion by increasing the proportion of the screen space occupied by the images. Similar effects for salience can be seen in the images in different versions of the story of *William Tell*. Figure 4.3, from an older humorous version of the story now available online (Wodehouse, 1904), shows Gessler ordering Tell to shoot the apple from his son's head. Figure 4.4 shows an image of the same episode from the picture book by Margaret Early (1991). There are also significant differences in the layout or compositional meanings of these images. In Figure 4.3 the participants take up most of the image space whereas in Figure 4.4 the participants are more dispersed and the background is given more space. Gessler is prominently located in the centre of Figure 4.3 and Tell's salience in the foreground is emphasized by the greater colour saturation of his hair and clothes. In Figure 4.3 there is not the same degree of salience afforded to the main characters.

The information value associated in Western culture with different positions on the page is also discussed by Kress and van Leeuwen (1996). One of these refers to the conventional role of information on the left-hand side of a page or screen as typically that which is 'given' or 'familiar' and that on the right-hand side as typically 'new' information. Not all pages or screens have this kind of horizontal polarization, but some of the online stories dealt with in this book exemplify this compositional resource. For example, in Figure 2.6 the tag for Wolstencroft is positioned on the right-hand side in the 'new' position, emphasizing that this is the novel aspect of this bear story. The online story of *Vasalisa* (Rock, nd), discussed in Chapter 5, involves the use of an illustrated navigational frame on the left-hand side of the screen and the story episodes hyperlinked to the chapter symbols in this frame appear on the right-hand side of the screen as indicated in Figure 5.6. Hence what is given or familiar is located on the left and the recurrent 'new' information on the right as each new episode appears. Similar compositional effects can be seen in *About Time* (Swigart, 2002), also discussed in Chapter 5.

Even this brief outline of key elements in Kress and van Leeuwen's grammar of visual design indicates the ways in which knowledge about the visual resources for constructing the different kinds of meanings in images can enhance the opportunities for learning in talk around texts and increase the potential for expanding young readers' interpretive reading practices.

Reading grammatically: engaging with the form of language in literary texts

If we are to help children understand some of the ways in which literary texts achieve their effects through language, we need to have access to the kind of grammatical description that will enable us to talk directly about language form and its use in constructing the various kinds of meanings at stake. Systemic functional grammar (SFG) provides the kind of resource that teachers, and ultimately children (G. Williams, 1998, 2000), can use to enhance their understanding of the relationship between the 'what' and the 'how' of literary texts. Space will not permit a detailed account of SFG for these purposes here – for an accessible, chapter length introduction to basic concepts of functional grammar see Ravelli (2000), Unsworth (2001) and Williams (1993) – however we can gain an appreciation of the use of key functional grammatical concepts as interpretive resources by applying them to a 'grammatical' reading of segments of works of e-literature discussed in more detail in the subsequent chapters.

SFG proposes that all clauses in all texts simultaneously construct three types of meanings. These are the same three types of meanings described earlier as the basis for Kress and van Leeuwen's (1996) grammar of visual design:

- ideational meanings involve the representation of objects, events and the circumstances of their relations in the material world;
- interpersonal meanings involve the nature of the relationships among the interactive participants;
- textual meanings deal with the ways in which linguistic signs can cohere to form texts.

Ideational meanings are realized grammatically by Participants, Processes and Circumstances of various kinds. Circumstances correspond to what is traditionally known as adverbs and adverbial phrases. Processes correspond to the verbal group, but Processes are semantically differentiated. Material Processes, for example, express actions or events (like 'walk', 'sit' or 'travel'), while Mental Processes deal with thinking, feeling and perceiving ('understand', 'detest', 'see') and Verbal Processes deal with saying of various kinds ('demand', 'shout', 'plead'). These different kinds of Processes have their own specific categories of participants. Material Processes, for example, entail Actors (those initiating the process) and Goals (those to whom the action is directed), while Mental Processes entail a Sensor and a Phenomenon.

Interpersonal meanings are realized by the mood and modality systems. Much of this aspect of functional grammar is familiar to those with experience of more traditional grammars. The declarative, for example, is realized by the ordering of Subject followed by the Finite Verb, while the interrogative is realized by the inversion of this ordering. The modal verbs (can, should, must etc.) indicate degrees of obligation or inclination and modal adverbs like 'frequently', 'probably', 'certainly' enable personal stance to be communicated through negotiating the semantic space between 'yes' and 'no'.

Textual meanings are realized in part by the Theme/Rheme system. The Theme of a clause in functional grammar is that element that is in first position. This is usually the Subject ('*The weather* is mild in Australia'), but an adverbial element could be placed in Theme position ('*In Australia* the weather is mild'). This is not as common, and is likely to occur more frequently in more formal written language than in informal spoken language. When some ideational element other the Subject is placed in first position in a clause, we call it a Marked Theme, in that it draws more attention due to its relatively less frequent use.

Although the three major grammatical systems realize the three main types of meaning simultaneously, in order to show how they can be used as a resource for understanding the patterned construction of literary texts, the construction of ideational, interpersonal and textual meanings will be discussed separately in the following three sub-sections.

Ideational meanings: processes, participants and circumstances

Ideational meanings are realized by the grammatical structures of the transitivity system involving Participants, Processes and Circumstances of various kinds. Primary school children who had very little prior systematic knowledge of grammar were able to use functional descriptions of Process types, Participant and Circumstances, to see how different examples encode a variety of meanings relevant to the representation of character in Anthony Browne's *Piggybook* (1986). The teacher drew attention to 'said' as the most common verb used in quoting or reporting speech and then experimented with the children in choosing alternative verbal processes like 'yelled', 'whispered' etc. It was this orientation and subsequent discrimination of verbal processes (saying verbs) that enabled the children to discuss the effect of

Browne's selection of verbal processes like 'squealed', 'grunted', 'snorted' and 'snuffled'. The children were also able to appreciate that in the first part of the story Mrs Piggott was the only Actor engaged in Material Processes (action verbs) that entailed a Goal. While Mr Piggott and the boys were Actors in Material Processes they didn't actually act upon anything whereas Mrs Piggott 'washed all the breakfast things … made all the beds … vacuumed all the carpets … and then she went to work.' At the end of the story however, all of the characters were Actors in Material Processes that had Goals, for example 'Patrick and Simon made the beds. Mr Piggott did the ironing' (G. Williams, 1998).

In Chapter 5 the role of the grammatical choices in constructing the main character in the story of *Wolstencroft the Bear* (K. Lewis and Moore, 2003) is described in the context of designing a learning activity for children to help them to understand this use of grammar as a resource for meaning. Most of the Processes or verbs Wolstencroft is associated with are Mental Processes or thinking and feeling verbs. In some cases Wolstencroft is the Sensor and the Phenomenon is a participant in the clause:

He	liked	its light, tinkling sounds.
Sensor	Process: Mental (affect)	Phenomenon

In many cases the Mental Processes of cognition, or thinking verbs, 'project' the actual thoughts of Wolstencroft as a second clause:

He knew
that tears would make his eyes all puffy and red.
He hoped
that Santa Claus would drop by on Christmas Eve and deliver him to a good home.

Because the majority of the verbs associated with Wolstencroft are these various forms of Mental Processes, we feel we are experiencing what is part of his consciousness and this aligns our sympathies very much with the character.

This kind of grammatical analysis also frequently assists in explicating the different narrative approaches taken in two versions of ostensibly the 'same' story. In Chapter 4, the CD-ROM version of *Mulan* (Disney, 1998) is compared with the version entitled *The Song of Mulan* (Lee, 1995). In the CD narrative the Processes are mostly Material or action verbs with very few Mental Processes or verbs of thinking, feeling or perception. In *The Song of Mulan* (Lee, 1995), however, there is a pattern of frequent use of the verb 'hear', indicating Mulan's reflection on her leaving home:

I	hear	my father's voice no more, only the rush of the river.
Sensor	Process: Mental	Phenomenon

Similarly, in Chapter 4, the episode from *William Tell Told Again* (Wodehouse, 1904) where Gessler orders Tell to shoot the apple from his son's head, is described as being much more actively told than is the same episode in the Margaret Early (1991) version of the story. This is reflected in the greater proportion of Material Processes or action verbs in the Wodehouse version.

I	should like to test	that feat of yours.
Actor	Process: Material	Goal

I	am going to put	this apple	on your son's head.
Actor	Process: Material	Goal	Circumstance: Location

But it is not only the greater proportion of Material Processes, but also that these Processes have Goals. That is, someone is doing something to someone or something else, whereas in the Early version there are fewer Material Processes and no Goals since the Actors are not acting on anyone or anything:

If	you	succeed
	Actor	Process: Material
	You both	shall go free
	Actor	Process: Material
But if	you	fail
	Actor	Process: Material

The importance as an element of narration of the writer's selection of processes for a participant to engage in has been noted by Knowles and Malmkjaer (1996). They illustrate this in their discussion of *The Secret Garden*, showing how the development of Mary Lennox from inactivity to action is traceable through a shift in the participant roles to which she is assigned. In chapter one, for example, 37.5 per cent of the instances in which she is mentioned is as Actor in material processes; 24.3 per cent as Goal; 20.1 per cent as Senser in mental processes and 13.8 per cent as Carrier in relational attributive processes. By chapter eight, however, Mary is Actor in material processes on 62.5 per cent of occasions in which she is mentioned. The pedagogical potential of explicit attention to the grammatical construction of characterization has been picked up in school syllabus documents. The Queensland English 1–10 Syllabus (Education Queensland, 1995), for example, provides sample lessons using these ideas to teach children about characterization in Judith Wright's short story *The Ant Lion* (quoted in Education Queensland, 1995). However, to this point we have considered only the grammatical construction of textual and ideational meaning. We will now consider how the grammatical construction of interpersonal meaning can also be seen as a resource for the interpretive reading of literary texts.

Interpersonal meaning: mood and modality

Interpersonal meaning concerns the nature of the relationships among the participants in the context of the particular text. These relationships are realized largely through the grammatical systems of mood and modality and other resources such as mode of address or 'vocatives'. We can see the effects of choices within the mood and modality systems in the nature and distribution

BLACKBURN COLLEGE LIBRARY

of statements, questions and commands among the participants. The two versions of the *William Tell* episode are discussed further in Chapter 4 from this perspective. In the Wodehouse text, Gessler makes two quite direct commands to Tell: '... take your bow' ... 'and get ready to shoot'. The only command in the Early text is more indirect in the 'Let' form 'Let me see you shoot an apple from your son's head'. In the Wodehouse version there are also a number of direct questions:

> Why was that? What did you intend to do with that arrow, Tell?
> ...
> Why was it?
> ...
> Why did you take out that second arrow?

whereas in the Early version there are no direct questions. The form of the single question in this text is a clause projected by the mental process (thinking verb) 'know':

> 'I am curious to know why you hid a second arrow in your jerkin ...'

The distinctive use of the vocatives referring to Gessler in the Wodehouse text: 'Your Excellency', 'Excellency' and 'My lord', suggesting an exaggerated and mocking subservience, is contrasted with the single vocative 'Tell' used by Gessler towards the archer. The more dialogic nature of the Wodehouse text achieved by such resources increases the immediacy and dramatic effect and enhances the more humorous approach.

The resources of modality include the modal verbs which construct high, medium or low values of obligation or inclination, such as 'could', 'should' and 'must', and modal adverbs, which construct similar meanings like 'possibly', 'probably', 'certainly' and 'maybe', 'perhaps' and 'of course'. Elsewhere (Unsworth, in press) I have described the differences in the final marriage scenes in the original *Shrek* storybook (Steig, 1990) and the more recent e-book (Ashby, 2001) based on the first Shrek movie. In the book, the modal adverbs and verbs build the intensity of the attraction and the imperative to marry. The modal adverb 'just' intensifies the attraction:

> Your lumpy nose, your pointy head,
> Your wicked eyes, so livid red,
> <u>Just</u> kill me.

The low modal verb 'could' followed by the repetition of the mental verb 'know' further indicates the extent of Shrek's attraction to the ugliness of the Princess:

> I could go on
> I know
> you know
> The reason why I *love* you so –
> You're *ugh-ly*!

And finally there is the higher modal verb used with 'marry' – 'we *should* marry'. By contrast, in the digital text the modal adverbs and verbs intensify the concern about not being perceived as beautiful:

> She was still an ogress
> ...
> I'm supposed to be beautiful

For further examples of the role of the analyses of the grammar of interpersonal meaning in understanding the construction of characterization in children's literature see Unsworth (2001; 2002).

Textual meanings: theme

One of the issues involved in textual meaning is how the information value of elements of the communication is indicated. In face-to-face communication these kinds of emphases can be achieved intonationally. In written English, a principal grammatical resource for this purpose is word order. Here we will consider how the writer indicates his/her point of departure, orienting the reader to what the clause is 'about'. In English this is achieved by location at the beginning of the clause and is referred to as the Theme in systemic functional grammar. The Topical Theme is the first participant, process or circumstance in the clause. In the following clauses from the Wodehouse version of *William Tell* (Wodehouse, 1904) the Topical Theme in each case is 'I' referring to Gessler:

Theme	Rheme
I	should like to test that feat of yours.
I	see you have it in your hand.
I	am going to put this apple on your son's head.

In the following clause from the Early version of the story, the Topical Theme is 'you both' referring to Tell and his son:

Theme	Rheme
You both	shall go free.

In Chapter 4 in Table 4.6 it is clear how much more the Wodehouse text constructs Gessler as the point of departure.

In dependent clauses the conjunction necessarily comes before the first participant, process or circumstance and hence it is referred to as a Textual Theme as shown in the following clauses from the Early text:

Textual Theme	Topical Theme	Rheme
If	you	are such a fine marksman ...
If	you	succeed ...
But if	you	fail ...

If there is a modal adverb or vocative at the beginning of a clause, it must be positioned after the Textual Theme and before the Topical Theme, as in this clause from the Wodehouse version of the *William Tell* episode:

Textual Theme	Interpersonal Theme	Topical Theme	Rheme
Now	Tell	I	have here an apple ...

In the examples above, all of the Topical Themes are the grammatical subject of the clause and this is the default option in English. If something other than the subject, such as a circumstance, appears first as Topical Theme in the clause, then this is referred to as a 'marked' Theme. Marked Themes are less frequent and consequently draw particular attention to what is in Theme position. In the e-book version of Shrek, marked Topical Themes emphasize time:

Theme (marked)	Rheme
Suddenly	the ogress swirled into the air.
Before long	Fiona and Shrek had their own wedding – in the swamp.

In the original story book there is only one Marked Theme and it emphasizes the close attraction of Shrek and the Princess:

Theme (marked)	Rheme
Like fire and smoke	these two belonged together.

The role of the pattern of Theme selection in highlighting readers' perception of the structure of narrative and drawing attention to experiential meanings integral to interpretive possibilities of text segments has recently been discussed by Knowles and Malmkjaer (1996) in relation to *The Secret Garden* (Burnett, 1992). They also discuss Thematization as 'a powerful tool for reinforcing a writer's explicit message'. To support this discussion they analyse part of the opening paragraph in the final chapter of *The Secret Garden* to point out how Burnett used Theme choice to pre-empt opposition to her strong claim about the effect of the mental lives of Mary and Colin on their physical condition.

Exploration of the patterning of Theme selection is an aspect of the analysis of textual meaning-making which can be introduced to very young readers as they begin to consider simple structural features of narrative. It can also introduce young readers to the ways in which writers draw attention to interpretive possibilities of story and can be extended to the investigation of quite complex issues about the textual construction of ideological positioning in narratives. However, the textual meanings realized by Theme selection also need to be seen in a complementary relation to the realization of ideational and interpersonal meanings.

Conclusion

The use of these kinds of functional grammatical analyses and related image analyses have been shown to be viable, engaging and productive in classroom work with children (Callow and Zammit, 2002; G. Williams, 1998, 2000) and have been incorporated in state education syllabus documents (Education Queensland, 1995; New South Wales Board of Studies, 1998). They have also been useful in explicating image/text relations in picture books, such as *The Rabbits* by John Marsden and Shaun Tan (Marsden and Tan, 1998), exemplifying aspects of Dresang's (Dresang, 1999; Dresang and McClelland, 1999) radical-change in books for children in a digital age (Unsworth and Wheeler, 2002). In the subsequent chapters, as one part of our exploration of e-literature for children, we will show how these analytical approaches to the language and images of literary narratives can provide enjoyable learning experiences, which will resource students in developing a critical understanding of both how and what texts teach.

Chapter 3

Learning through web contexts of book-based literary narratives

Introduction

In this chapter we are concerned with the ways in which information and communication technology (ICT), especially through the pervasiveness of the world wide web (www), is more and more becoming a routine aspect of engaging with reading literature in book form. This kind of electronic enhancement of literary narrative is expanding and altering the nature of the 'possible worlds' we have traditionally thought of as being contained within the covers of the books – and it is also changing the ways we respond to these story worlds. These expanded dimensions of the experience of story are a significant part of what encourages many young readers to maintain their engagement with extended and intensive reading of books in a multimedia world. Learning experiences derived from this richly electronically enhanced ecology of today's literature for children and young people are an important resource in developing their literary understanding. This chapter will describe a variety of contexts of electronic enhancement of recently published picture books and novels as a basis for generating classroom learning experiences that will optimize the survival of engaged reading of such books by children of the internet age.

Context of composition

In the process of composing a story, authors frequently construct a range of episodes and events and describe various backgrounds and artefacts, some of which are eventually discarded as they hone the story towards its published version. Notes on factual events, sometimes in authors own lives, may form a part of the initial basis of the story composition, and these may later be abandoned, adapted, disguised or absorbed into the fabric of the narrative.

These dimensions of story are what Margaret Mackey discusses as the 'phase space':

> things that might have happened in the plot but did not, aspects of characters or incidents that are known to the author or that can be imagined by readers but that are not laid down in the novel itself.
>
> (Mackey, 1999: 20)

Mackey goes on to point out that the 'phase space' is frequently targeted by teachers who invite students to:

> write a 'missing' chapter, include a scene that occurs offstage from the main action of the novel, or supply school report cards or news bulletins relating to characters or incidents of a book.
>
> (Mackey, 1999: 20)

Aspects of the phase space of stories are now increasingly publicized by authors as significant elements of the story-world, which is implicated but not contained by the book. Authors' websites and authors' input on publishers' websites are a prime means of this kind of story enhancement. Mackey (1999) draws attention to the website sponsored by Random House, the American publishers of Philip Pullman's trilogy *His Dark Materials*:

> Each book has its own site, and each site offers background information about the story world that does not appear in the novels but manifestly applies to them. The *Golden Compass* site offers a detailed history and symbol glossary of the alethiometer, the truth-telling compass of the American title. On the *Subtle Knife* site, there is information about *Liber Angelorum*, the book of the angels, which now, we are told, survives only in the form of evocative 'scraps' saved after a fire in the Torre degli Angeli.
>
> (Mackey, 1999: 18)

A recent addition to this site (http://www.randomhouse.com/features/pullman/index.html) is a new book by Philip Pullman entitled *Lyra's Oxford*, set in the world of the *His Dark Materials* saga.

The J.K. Rowling official website (http://www.jkrowling.com/) contains a great deal of phase space material about the Harry Potter books. Under 'extra stuff' she notes:

> Here are a few bits and pieces from my notes that you might find
> interesting; some scenes that were cut, a few extra details about some of
> the characters ...

In the 'edits' section of 'extra stuff' Rowling discusses a number of editorial
variations across the English and American versions of *The Philosopher's
Stone* (Rowling, 1997) and the movie version, especially about the background
of Thomas Dean. She also discusses the 'many different versions of the first
chapter of *The Philosopher's Stone*' and how they relate to the final published
version. In the early versions '… you actually saw Voldemort entering Godric's
Hollow and killing the Potters'.

> Other drafts included a character by the name of 'Pyrites' ... He was a
> servant of Voldemort and was meeting Sirius in front of the Potter's
> house.

Similarly, information about the early drafting of characters is provided.
For example, Rowling discusses the modelling of Gilderoy Lockhart on
someone she knew and how Ron often behaves like her oldest friend Sean.
She also provides Nearly Headless Nick's account of his near decapitation
'in his own moving words'. In addition, the site provides an extensive
illustrated biography, and amongst other links such as fan sites, publishers'
sites, and the Warner Brothers Harry Potter site, there is a kind of anticipatory
phase space link to 'rumours' about the title of the sixth book.

Authors are often asked about the genesis of their stories and consequently
their websites now provide information about their negotiation of the phase
space in their writing. David Almond (http://www.davidalmond.com/)
comments on the origin of *Skellig* (Almond, 1998), explaining that the name
comes from the Skellig Islands, which are off the south west coast of Ireland.
However, Almond says that he is not sure who or what the character Skellig
is or where he comes from, how he got into the garage or where he goes to
in the end. 'He remains a mystery – like much of life.' Aiden Chambers'
website (http://www.aidanchambers.co.uk/) provides some information on
the genesis of *The Present Takers* (Chambers, 1983). Chambers says that a
friend found out that his daughter was being badly bullied at school and that
no one seemed to be able to stop it. The friend asked Aiden Chambers – as
an ex-teacher – what he would do. Chambers says he didn't know what to
say so he wrote the book to find an answer. He indicates that the first half of

the story is pretty much the way it happened in real life and that he made up the second half.

In addition to authors' and publishers' websites, phase space information is sometimes included in sites offering resources for teachers. For example, some interesting material on Stephen Crane's *The Red Badge of Courage* (1974) is provided on the Discover School site (http://www.school.discovery. com/lessonplans/programs/redbadge/). This includes factual information on the battle at Chancellorsville, which, according to the website, Crane indicated in *The Veteran* was the site of the battle in *The Red Badge of Courage*. The Discover School site provides similar useful links relating to range of fiction for children and adolescents. The page dealing with H. G. Wells' *The War of the Worlds* (1986), for example, provides links to the original famed Orson Wells radio broadcast of 1938 in the USA, which apparently set in motion a genuine mass panic (http://www.war-ofthe-worlds.co.uk/).

An interesting phase space exploration for students based on Robert O'Brien's *Z for Zachariah* (1998) is found on the Read-Write-Think web-based resources site developed by the International Reading Association and the National Council for the Teaching of English (http://www.readwritethink. org/lessons/lesson_view.asp?id=5). Throughout this novel the narrator, Ann Burden, is faced with a number of difficult decisions as she struggles to survive in a post-nuclear holocaust world. The students are asked to identify these decision points and rewrite the remaining portion of the plot based around the predicted effect of Ann's making the opposite choice from that recorded in the novel. For example, what if Ann had killed Faro during chapter 23 instead of his dying later in the story? The detailed online lesson plan includes careful scaffolding of students towards this task and incorporates free online flowcharting software ('Plot Alternatives Designer') to assist students to map the cause-effect chains resulting from different key decision choices made by the narrator.

As well as examining aspects of the phase space of stories provided by authors, publishers and educators, and imagining the plot ramifications of alternative decisions by characters, students might be encouraged to project the authors' own information about what was discarded from the published narrative and 're-vision' the stories to construct versions in which these elements are retained. What might have been the role of Pyrites in *The Philosopher's Stone*? How could Nearly Headless Nick's own account of his mishap be interpolated into the story? What if Hermione's last name was Puckle, as originally intended, rather than Granger?

The context of invitation

The promotion of the book and enticement of potential readers on publishers' and authors' websites frequently includes excerpts from the book with audio to enable the potential reader to hear the story segment read aloud – sometimes by the author. David Almond on his website (http://www.davidalmond.com/) reads short excerpts from his novels. One chapter of each of Morris Gleitzman's books is available in text form on his site (http://www.morrisgleitzman.com/) and audio is available for several of these. Gleitzman also offers hardcopy books as 'giveaways' and a similar offer is made from time to time by Paul Jennings (http://www.pauljennings.com.au/).

Readers may sample electronically one chapter from each of Beverly Cleary's books online (http://www.beverlycleary.com) and similarly they can read a sample chapter from the works of many authors including Philip Pullman's *The Golden Compass, The Subtle Knife* and *The Amber Spyglass* on the Random House site (http://www.randomhouse.com/features/pullman/index.html). Online booksellers like Amazon.com (http://www.amazon.com) and Alibris (http://www.alibris.com) also include short segments of the books with images (but not usually with audio), and similar opportunities are available on many publishers' sites, such as Pan Asian Publications' beautiful online presentation (http://www.panap.com/mulan_order.html) of the early episodes of *The Ballad of Mulan* (Nan Zhang, 1998).

While a number of students may engage informally with such invitational activities from time to time, it is also possible to recruit these resources for systematic classroom use. Students might explore the sample chapters of books they will pursue for either close or wide reading in the classroom program. They could be asked to rate the effectiveness of the 'teaser' chapters in enticing them to read the whole book. This rating might be based on a number of features of the narrative including aspects of the audio presentation where relevant. Students might undertake the task in small groups based on a common book choice. After individually experiencing and rating the chapters, they could discuss the variation and commonality of their responses as a group. The next move might be to re-group the students to form new groups consisting of one member of each of the original groups. They could then discuss the relative effectiveness of the teaser chapters of different books.

A relatively recent form of invitation or enticement to buy and read new books is the inclusion of electronic games related to the story on publishers' and/or authors' websites. Such games are beginning to be incorporated more commonly as part of the post-reading experience, but they are also becoming part of a suite of experiences comprising the initial introduction of new narratives. Paul Jennings and Morris Gleitzman now promote their *Wicked* stories (Jennings and Gleitzman, 1998) through online games. On Paul Jennings' website (http://www.pauljennings.com.au/) he now asks visitors:

> Why don't you check out our 'Wicked' animations? Here you can meet Rory, Dawn and the Appleman or have a go at squashing the SLOBBERERS.

Jon Scieszka and Lane Smith are probably most well-known for their metafictive work, *The Stinky Cheese Man and Other Fairly Stupid Tales* (Scieszka and Smith, 1992). Their website (http://www.baloneyhenryp.com/) currently promotes their new book, *Henry P. Baloney,* and included on the site is an electronic game entitled 'Henry's Piskas Game'. Henry P. Baloney is a small being from outer space, who has difficulty with being late for school and with learning. He has to use his imagination to get out of difficult situations and to manage the problems that confront him. The online game instructions indicate that

> Henry has landed on the planet Astrosus and is about to be eaten by the Astro Guys! To save himself he entertains the Astro Guys by drawing Piskas. Can you guess what Henry is drawing? Click on the correct answer before Henry finishes his Piskas.

The idea of using online games to entice readers is not confined to authors of children's books. The newspaper *Sydney Morning Herald* carried a story by Alan Mascarenhas in its weekend edition for July 17–18, 2004, about author Bruno Bouchet who designed an online game to sell his book, *French Letters*, a satire on the phenomenon of the fabled 'sea change' – specifically baby boomers who descend on villages in the South of France to sort out their mid-life crises.

As online games become more commonly used to attract young readers to books, classroom activities involving children in rating their effectiveness for this purpose could also be undertaken in a similar manner to that suggested for rating teaser chapters. It would also be useful then to have the children who read the book revisit the games and discuss again how effective they were in predicting the appeal of the book and foreshadowing the key events and aspects of characterization.

While the sample story segments, often with online audio and the promotional online games are the kinds of 'teasers' that young readers are likely to find most inviting, it is also significant to note the prominence given to reviews by online booksellers. Some of the reviews are provided by 'in house' staff, but many are purportedly provided by customers who have read the books. The booksellers apparently value the reviews as selling strategies since those who contribute are often rewarded by book discounts or credits. Again while some children may informally engage in reading and/ or writing and submitting such reviews, this phenomenon might also be recruited for systematic classroom purposes. Here it is important to address the development of students' capacity for critical reading of such reviews and also to model writing and then scaffold their composition of reviews for submission. This is a significant challenge since the reviews by professional adults that appear in hard copy journals and online frequently adopt a very superficial approach to the narrative art of children's books. For a further discussion of this issue and guidelines and examples of scaffolded teaching and classroom work with reviews of picture books see Unsworth and Wheeler (2002).

The context of appreciation

Fan sites for J. K. Rowling and the Harry Potter books might well have been expected, and indeed there is a plethora of these, from which those considered most impressive achieve recognition on the official J. K. Rowling site (http:// www.jkrowling.com/). One such fan site is Mugglenet (http://www.mugglenet. com/) managed by 17-year-old webmaster Emerson. The site includes very wide ranging responses to, and information and speculation about, the current and forthcoming books, videos, games and other artefacts related to the stories. Visitors to the site can participate in forums, polls and competitions, and contribute to editorials and submit artwork as well as engage in a wide range of other fan activities. However, fans of many other authors construct sites to celebrate the work of their favourite writers. The Australian author Isobelle Carmody is the subject of several such sites. Carmody is best known for her strikingly innovative fantasy fiction books. These include the highly-acclaimed *Obernewtyn* novels, *Obernewtyn* (1987), *The Farseekers* (1990) and *Ashling* (1996). She has also written several short stories, stand-alone stories such as *Scatterlings* (1992), and *The Gathering* (1993), and a recent illustrated story *Dreamwalker* with Steven Woolman (2001). Some of the

Carmody fan sites are fairly simple, inviting contributions of reader reviews of the books and participation in an online discussion forum (http://www.allscifi.com/Topics/Topic_98.asp), while others focus more on the site owner's celebration of the author rather than sharing of responses to the work, such as the *Enchanted Forest* (http://www.geocities.com/EnchantedForest/Cove/1933/) site, which includes a biography of the author, review of the books and links to other sites concerned with Carmody's writing. A much more elaborate site is conducted by the members of the obernewtyn.net club (http://www.obernewtyn.net/). This site has online message board discussions about the books, a frequently asked questions section, as well as interviews with the author and a host of fan activities. These include the collaborative construction of an online story in the vein of Carmody's work, fan fiction competitions, contributions of fan art, as well as opportunities to meet at conferences in the real world off-line. Membership is free and participants can choose to join a number of 'guilds' named in accordance with aspects of the Carmody fictional worlds (Ashlings, Dreamweavers, Mystics or Wanderers).

A different kind of fan site is one devoted to the story of *The Little Prince* by Antoine de Saint-Exupery (http://members.lycos.nl/tlp/). As well as a celebration of the book, and biographical information about the author, this avid fan displays images of his vast collection of the many editions of the book – in many different languages – of which he has several copies, and offers to provide a two for one swap to anyone who can provide him with an edition that he does not have. The site also provides links to a multitude of online versions of the story, again in many different languages.

Similar celebratory sites can be found for many other authors such as 'A Tribute to William Golding' (http://www.geocities.com/Athens/Forum/6249), an 'excellence' page for Gillian Rubinstein (http://www.carnelianvalley.com/hearn/) and James Bow's testimony to a well-known Australian children's author: 'If I had to make a list of authors who have most influenced me, one of the higher ones on my list would have to be Patricia Wrightson' (http://www.bowjamesbow.net/2004/01/12-were_back.shtml). Students can be encouraged to locate such sites for their favourite authors and discuss the ways in which these highlight the aspects of the writers' work that they find engaging. If children re-group they can share their findings so that others become aware of fan sites for authors they are less familiar with. This may well lead to their expanding the range of authors in their reading repertoire and enriching their appreciation and understanding of the works.

The context of interpretation

There are two main kinds of web-based contexts for facilitating young readers' interpretive practices in reading literary novels and picture books. The first are lesson plans and learning activities of various kinds posted on publishers' and educators' sites. The second are online options for readers to communicate and discuss with others their thinking about the books they are reading. These include emailing the author (or at least the manager of the author's email/website) and chat rooms and online forums of various kinds, as well as student constructed websites displaying their work with particular books. Online lesson plans and activities will be dealt with briefly since most of these do not involve students in using the affordances of computer-technology and the web, and simply convey fairly traditional 'pen and paper' tasks. Then the variety of options for online communication and discussion and the learning experiences they facilitate will be outlined.

Online lesson plans and learning activities

There are many sites which offer classroom teaching resources for books as '.pdf' files that can be downloaded free of charge. Typical is the Beverly Cleary site (http://www.beverlycleary.com). The 'downloadable' materials for the 'Ramona' books (Cleary, 1976, 1978, 1981, 1982, 1984, 1986) include summaries of the stories, character sketches of Ramona, her family, friends, teachers and other characters, and suggestions for classroom use of the books, such as 'read alouds', 'independent reading' and 'literature circles'. Lothian books (http://www.lothian.com.au) provide similar 'downloadable' teaching resources for many books including *The Rabbits* (Marsden and Tan, 1998), which won the Children's Book Council Picturebook of the Year award in Australia in 1999, and *Dreamwalker* (Carmody and Woolman, 2001). An anthology of sites offering such resources for children's literature for the primary grades has been compiled by Marilyn Newman (2004). Some of these kinds of online resources provide engaging and challenging learning experiences but they typically do not involve the use of technology or the wider resources of the web. There are, of course, some notable exceptions such as the *Z for Zachariah* site noted above and the Discover School site (http://www.school.discovery.com/lessonplans/programs/redbadge/), which provide links to short online video clips that can be viewed free of charge with the free downloadable 'Real Player'. On this site links are provided to such video clips for novels like *The Red Badge of Courage* (Crane, 1974),

The Lord of the Flies (Golding, 1963), *Gulliver's Travels* (Swift, 1972) and *The Legend of King Arthur* (Malory, 1975). The free video clips are intended to encourage teachers to purchase the full video (which they can do online) and use this in conjunction with the learning experiences outlined on the site. However, the gratis clips themselves are a useful resource for classroom work. As earlier suggested in relation to online teaser sample chapters and promotional games, students can rate the clips in terms of their introduction to reflection of various aspects of the novel and compare their findings within and across small groups within the class dealing with the same or different novels/clips.

Opportunities for online communication and discussion about books

Many author websites encourage readers to communicate with the author by email. In most cases this does not mean direct email contact, but emails are frequently responded to by the author's posting of general responses to common email topics. This is the case with the Morris Gleitzman site referred to earlier. Some sites solicit children's reviews of books, which can be typed on screen on the website and submitted electronically, as is the case on the Roald Dahl site (http://www.roalddahl.com/index3.htm). Specialist children's literature websites (http://www.acs.ucalgary.ca/~dkbrown/) include links to online discussion groups for children where they can submit reviews of books and read reviews of others. Such online forums and discussion groups are also a feature of many of the fan sites such as those noted above. Some government education authorities and independent school organizations sponsor online forums that are more integrated into children's school experience of literature. One very successful such forum is located on the Netlibris site in Finland (http://www.netlibris.net/international/), and a related but somewhat different approach is found in the range of online 'book raps' projects.

The Netlibris site (http://www.netlibris.net/international/) was designed by Finnish educators Minttu Ollila and Teresa Volotinen in 1996 to enrich teacher's programs in children's literature. The initial name of the project was *Matilda* (from the title of the Roald Dahl book) and was oriented to children in the primary school. It was originally designed to provide an ICT environment that appealed particularly to girls and to encourage and challenge young readers to extend the number and variety of literary texts they read, and to share their reading experience with other keen readers. It was also intended to encourage collaborative, independent learning and to provide opportunities for children to play a significant role in designing and evaluating their own learning, as well as enriching the regular classroom experience of children's literature. The project was extremely successful, winning many local and international awards for excellence in the use of ICT in the classroom and learning. Not only did it achieve the aims of engaging the interest of girls in an ICT environment, but it also increased the interest of boys in reading literature. Consequently, the site has flourished and extended its programs to include all levels in the Finnish school system. The site reaches well beyond Finland and is now an international program where teachers and children from many different countries talk about children's literature. For example, Dr Angela Thomas, at the University of Sydney in Australia, has facilitated a joint project between an Australian and a Finnish teacher of children in the upper primary school. Initially the Australian and Finnish children interacted via the Netlibris site. Both groups of children read the David Almond novel *Skellig* (Almond, 1998). This award-winning novel was popular with both audiences and was accessible in both English and Finnish. As children in each class learnt progressively more about the novel, they were encouraged to write their thoughts in a formal response to the text, post it to Netlibris, and interact with the ideas posted by other children. To follow up this work the Finnish and Australian groups used the ELLIE website (http://sirius.linknet.com.au/ellie/) – Electronic Literature and Literacies in International Education – designed by Thomas to foster educational work with new forms of electronic literature for children. The e-text selected was originally published by the BBC to accompany a television series, and was titled *Spywatch*. This story is set in wartime England and traces the activities of children who are investigating the townspeople to help the police discover a spy who is active in the town. Since the text was not extensive and a good deal of the narrative was constructed by the images, the Finnish children were able to negotiate the story in English. Much of the discussion was about the nature and role of the images. (The BBC have since taken the e-book offline but educator Ben Clarke has (with permission) made it available for free download for teachers at: http://www.lookandread. fsnet.co.uk/ downloads/sites.html#spy.) Further discussion of the Netlibris project can be found in the online paper by Miller (2000).

Book raps are essentially a teacher managed online discussion forum organized via email to which children post responses to teacher-generated points for discussion about specified books (Simpson, 2004). Naming the activity 'rap' indicates the intention to distinguish this online work with literature from more traditional classroom approaches and to foster more open discussion among a wider range of readers. In Australia, rap sites are conducted by the Queensland University of Technology (QUT) (http://rite.ed.qut.edu.au/old_oz-teachernet/projects/book-rap/) and the New South Wales Department of Education and Training (NSW DET) (http://www.schools.nsw.edu.au/schoollibraries/teaching/raps/).

The Book Chat site in New Zealand (http://english.unitecnology.ac.nz/bookchat/home.html) is quite similar to the book raps approach and another somewhat similar site is the E-pal project in the USA (http://www.brick.net/~classact/nbooks.htm).

The personnel involved in designing and implementing book raps are the coordinator of the rap for a particular book, the webmaster who manages the postings on the host site and the teachers who implement the rap activities in their classrooms. Each rap focuses on one story and has its own coordinator. Successive raps may have different coordinators. The coordinator is a volunteer who creates and manages the rap. S/he chooses the book, generates the questions and activities the children will engage in, determines whether to provide links to author's sites and/or other relevant sites and formulates a series of questions, or rap points, that will guide the students through their interaction with the text. S/he also writes an inviting introduction to the rap to enthuse the children about participating in the activities, introducing herself as the rap coordinator. S/he then decides on the duration of the rap and the period during which it will be run, and then sends all of this information to the webmaster who reviews the material, provides guidance on any revisions that are necessary and eventually posts the rap information on the website. Summary information on forthcoming raps is advertised well in advance of their start dates to enable teachers to select the rap that they wish to join in time to plan for the incorporation of the rap into their usual work with literature in the classroom.

The book raps are organized by using two separate email lists. One list is available only to teachers as participants in or observers of the raps, the coordinator and the webmaster. This is essentially to provide support for teachers participating in the raps and to allow teachers to share ideas about implementation. The student list is where the information from the rap coordinator is communicated to the children – from the initial introduction and the rap points to the feedback provided on the rap point discussion and finally the rap 'wrap up'. It is on this list, of course, that students respond to the rap points, and they may also respond to the other students' postings. Participation in rap point discussion postings may be on an individual, small group or whole class basis. The book rap website includes links to the various current raps and also links to records of past raps. While these remain 'live' they are an interesting and useful resource for others as well as a fascinating retrospective for those who actually generated the rap responses.

Schools are increasingly facilitating the opportunity for students to publish their classroom literature work on the web. Since we have previously noted web resources for the novel Z for Zachariah (O'Brien, 1998), we can also note student produced websites about the book. One group of students from the Ulricianum high school in Germany produced a site (http://home.t-online.de/home/familie-bode/zfz/main.html) which contains information about the author, different covers for the novel, student summaries of the story and their opinions of it, discussions of the characters and the ending of the story (including an alternative ending), and a discussion linking aspects of the novel to the Bible. Visitors are invited to respond to the site by email and the students indicate that 'So many people have already visited this site'. They also provide a link to another Z for Zachariah site constructed by students in Switzerland (http://www.kl.unibe.ch/sec2/neufeld/arbeiten/mng/zachari/parent.html). This is a visually and rhetorically impressive site. It also provides a biography of the author, a grid showing story time and plot developments, character summaries, an alternative cover design and paintings inspired by the book, as well as an imaginary newspaper report of the devastation that constitutes the setting for the story.

School facilitated website construction by children showing their interpretive engagement with literature takes a variety of forms. At Dalton elementary school in New York, for several years a number of successive classes of fourth grade students have worked with their teacher Monica Edinger on two different types of projects with Lewis Carroll's *Alice's Adventures in Wonderland* (http://intranet.dalton.org/ms/alice/alice.html). The first project involved children in re-illustrating the story and publishing their new version online. The teacher modelled the illustrating process for the first two chapters and then the children took on the work for the remainder of the book. The steps are summarized by Lindsay, who was in Monica Edinger's grade four class at the time:

1. Ms. Edinger read the book Alice's Adventures in Wonderland to us.

2. We chose chapters (2 people to a chapter) to illustrate.
3. We drew pictures of the text we selected.
4. Ms. Feldman scanned our pictures into the computer.
5. We were told to write an essay and annotate the text we chose.
6. Ms. Gumport helped us put our essays on the site.

In the winter of 2003 Ms Edinger worked on the second type of project with a subsequent fourth grade class:

After a deep immersion into the story, its author, and more, the class created a toy theater production of the book. In teams, the children each wrote adaptations of the book's chapters; created characters, props, and scenery; practiced their plays, and performed them for each other. These were then videotaped and placed on student-made web pages for others to see and enjoy. This is the third year a class has done this project and this year's class was able to use last year's projects as models.

The products of both projects can be viewed on the Dalton site (http://intranet.dalton.org/ms/alice/alice.html). The approach taken by Monica Edinger can be modified and extended using other 'out of copyright' texts and other innovative scenarios that will capture children's interest and enhance their engagement with literary learning. The advent of new, accessible, computer software, enabling more innovative teaching and learning activities, can greatly assist teachers in generating these collaborative web-based learning environments. For example, Desktop Author 3 (http://www.desktopauthor.com) allows users to translate the book metaphor into digital form, so that screen-based story presentations are shown in book format with 'turnable' pages etc. Hyperlinks can be included and stories can be published to the web, distributed on floppy disks or made available as downloads from a website.

It is arguably the case that electronic forms of personal communication, favoured by and accessible to an increasing proportion of young people, can be highly facilitative of their intensified participation in communities of readers of literary works. Evolving forms of such individually accessible digital communication modes are making it easier to share multimodal textual information and exchange ideas about these texts with others on the other side of the world (or the other side of the street) almost instantly in cyberspace. For example, 'blogging' is a relatively recent form of communication that is rapidly being adopted by increasing numbers of young people. A 'blog' (abbreviated from 'web log') is an online facility for personal interactive commentary. Blogs resemble web pages, written by the page's editor ('blogger') expousing his/her ideas about any subject they like. One of the attractive features of blogging is that the pages are so simple and easy to generate and manage. The tools to make the blogs are located on a number of websites, perhaps the most well-known and popular of which is http://www.blogger.com – and they are free. Potential bloggers are required to enter their email addresses, select a user name and a password and provide a few other pieces of information. Then they can select a template for their blog, which determines what it will look like on the screen, and they are ready to type up their comments and publish them on the web. Blog entries can be very short or as long as one wishes. They can consist of text only or of text and images and sound bytes. Bloggers can list their interests in the personal profile section and then by clicking on particular interests, a list of other bloggers who have also listed that interest will be displayed. It is then possible to visit their blogs to explore their postings in this interest area. The potential for engaging children in learning via this facility are very exciting. Classroom communities might initially discuss listing literary works they are currently dealing with as common interests on children's individual blogs – such as 'children's literature' and 'Anthony Browne's picturebooks'. Students (and their teachers) could then exchange their ideas via their blogs. Of course this might be extended across classes, schools and countries – and many unplanned exchanges might occur. What is important is that children's involvement with ICT is able to be envisioned by teachers as facilitative of, rather than antithetical to, their sustained engagement with literary texts.

The context of adjunct-composition

One of the online contexts of literary texts which provide a great deal of enjoyable engagement in the story world created by the narrative is that of adjunct-composition, where readers become writers, developing new, alternative or modified episodes and story elements within the phase space of the original story. This can occur through collaborative online story composition and through the composition of 'fan fiction', as well as through the online multimodal, virtual interactive spaces know as story palaces. In this section we will briefly revisit the obernewtyn.net club (http://www.obernewtyn.net/) to look at their provision for collaborative online storying and fan fiction composition and then we will outline the role of some story palaces in providing opportunities for multimodal re-composition, focusing particularly on the Middle Earth site (http://www. middleearthpalace. com/palace.html).

The obernewtyn.net club (http://www.obernewtyn.net/) is a fan site for Australian author, Isobelle Carmody. Club members are invited to join a collaborative enterprise in writing a story in the vein of the Carmody novels. Potential contributors can read the story to date online, access character descriptions of the characters created thus far and visit the forum to add their contributions. The site also conducts fan fiction writing competitions and launched their eighth competition in May, 2004. One category of the competition is '*Obernewtyn* related writing (this may be something like a poem about a character, or an essay on which lands you think are spoken of in the chronicles, or some fan fiction etc.)'. Results of the competition and the winning pieces are published on the site. There is also provision in the Obernewtyn Multimedia Gallery (http://www.obernewtyn.net/multimedia/ menu.html) for visual 'adjunct-composition' of elements of the Carmody story worlds. Contributions displayed include paintings and drawings inspired by the stories and computer art like animated 'gifs' of story titles and downloadable screensavers such as 'Elixa's Obernewtyn Chronicles Quotes Screensaver'. Another impressive site that provides for adjunct-composition activities around the Carmody books is run by Jacqui (http://www.carmody-online.com/html/index.php).

The Palace is a visual virtual world where multiple users are able to create visual representations of characters and communicate meanings via these characters using both text and the manipulation and movement of the visual representations. A room in a palace looks something like a comic book backdrop into which participants locate the visual representations of characters (avatars) they have selected (or constructed through a form of computer programming language). Avatars can be attractive human figures, cartoon figures, accessories like sunglasses or objects like a guitar or a cloud. Each participant can select from a repertoire of avatars to represent different physical positions and attitudes. Some participants maintain a consistent selection of avatars while others tend to vary these quite often during any one online session. Palaces, then, are a multimodal interactive virtual space where many participants can interact to discuss various topics or collaboratively construct fictional worlds. Sometimes palaces are used as sites for composing 'fan fiction' where participants write alternative episodes for their favourite television shows or sequels for their favourite movies, adopt character roles and organize with other palace participants virtual performances of their creations, but palaces are also sites for such creative adjunct-composition of the fictional worlds of literary texts. A great deal of work in explicating the educational significance of palaces for young people has been done by Dr. Angela Thomas at the University of Sydney and this account draws significantly on her material (Thomas, 2000, 2001, 2004a, 2004b).

Some palaces can be accessed from inside a web browser, but to enable full use of all of the palace features, it is advisable to download the free *Palace User* software (http://www.palaceplanet.net). The construct of the palace environment incorporates a hierarchical differentiation of users at the levels of: member, wizard and god. Members can enter the palace, deploy avatars and participate in textual interaction, but wizards have access to regulatory commands. For example, they can execute commands that will prevent a participant from remaining on the site if such participants violate the agreed upon code of conduct, and wizards can extend the palace by adding rooms. Ultimate control, including the investing of wizard status, remains with the owner(s) of the site – god(s). Some new learning of the specialist language of palace commands which enable communication via text and avatar manipulation is required, but most palace sites refer users to online tutorials that enable them to learn sufficient of this to readily participate in the enjoyment of interacting in these virtual worlds.

Not surprisingly, there are a significant number of palace sites devoted to the world of Harry Potter, such as *Hogwarts* (Maykitten, 2004), *Harry Potters* (Aurora, 2004), and *Bloody Brilliant* (Layke, 2004). The Tolkien stories are also the focus for a number of Palace sites. A very appropriate g-rated example for children (providing a 'safe' online environment) is the Middle Earth palace (http://www.middleearthpalace.com/palace.html), which can be readily accessed via a web browser like Netscape or Internet Explorer (as well as the free downloadable specialist palace software). Instructions and free software required for joining this site are outlined in detail at the Middle Earth palace website. The owner of the palace is Laurie Sorenson, who adopts the name 'Nimue' in the palace world. Participants in the Middle Earth community come from many different countries and range in age from 11 to 70 years and have adopted names such as 'Elrond' and 'Hobbitness'. Tolkien's magic lands are depicted by beautiful artwork in the palace, which includes sound effects from movie versions of the books. Avatars of hobbits, elves and other inhabitants of Middle Earth are located in various rooms for children to find during their visits to the palace. The Middle Earth community members have their own language, called 'Elvish', but translations into English can always be obtained for those not proficient in Elvish. A key element of the Middle Earth palace is fantasy role-playing site, which enables children to participate in the imaginative recomposition of story elements from the phase space of the Tolkien stories. Another important feature of this site is the 'scaffolding' or support provided to novice users with drop down help menus, covering basics like movement from room to room in the palace, and a map showing where you are located in Middle Earth.

The variety of forms of online adjunct-composition around the experience of literary texts, the relative ease with which it is possible to participate in such activities, the enjoyment of the creation and community participation, the satisfying quality of the content and appearance of the material products of this collaboration, and the educational challenges and new forms of online mentoring and social learning that take place, are very compelling reasons for teachers to explore these contexts with children as opportunities for robust intellectual engagement in understanding new forms to literary experience as:

> Electronic media are not simply changing the way we tell stories: they are changing the very nature of story, of what we understand (or do not understand) to be narratives.
>
> (Hunt, 2000:111)

Developing knowledge about children's literature on the web

There is a veritable plethora of websites that can be used to support teaching with children's literature. To find sites at the point of need in teaching preparation one can, of course, deploy a 'search engine' like yahoo or google and search on the book title and/or author. This is likely to yield thousands of 'hits', which can then be refined by a more advanced search. However, it is useful to be aware of key categories of sites, which may enable more strategic 'surfing' to locate lesson support materials. Specialist sites dealing in detail with children's literature and education have become established over a number of years. Well regarded among these are the Carol Hurst site http://www.carolhurst.com/, the University of Calgary children's literature website http://www.acs.ucalgary.ca/~dkbrown/ and the Vandergrift's children's literature page hosted at Rutgers State University of New Jersey http://www.scils.rutgers.edu/~kvander/ChildrenLit/index.html. These sites provide access to a wealth of information including links to author websites, resources for classroom learning activities, critical discussion of issues in children's literature, links to professional journals and associations and other websites related to children's literature. Publishers' websites are often a good starting point, especially for established authors and/or illustrators who tend to work with the one publishing company. They typically provide a link to, or maintain, the author's/illustrator's own website. A number of the sites listed have links related to 'visual literacy' and some specialist sites focus on the work of illustrators of children's literature, such as the site developed by the society of book illustrators in Australia (http://www.thestylefile.com/about_us_page.htm). The International Board on Books for Young people (IBBY) (http://www.ibby. org) and the Children's Book Council (http://www.cbcbooks.org/) provide an international perspective on children's literature, including awards and prizes for outstanding literary works for young readers. A very significant resource for teachers in locating relevant websites and using web technology is the knowledge and experience of ICT that an increasing proportion of their students can contribute. An acknowledgement of the ways in which exponentially expanding and improving technology is changing the dynamics of pedagogic practices is essential to maintaining children's engagement with learning through literary texts – and technology.

Chapter 4

Classic and contemporary children's literature in electronic formats

Introduction

The re-presentation of children's books on CD-ROM and the web provides more opportunities for readers to experience stories they like in different media. It also creates more opportunities to learn about the ways in which variations in the use of language and images, as well as electronic features of hyperlinks and animation, can construct different interpretations of ostensibly the same story. In this chapter we will compare original book and multimedia electronic versions of well-known picture books and more extended classic and contemporary illustrated stories for children. Such comparisons can be the basis for classroom work designed to extend children's understanding of narrative techniques, their knowledge of verbal and visual grammar as a resource for meaning, and their critical comprehension of, and response to, literary texts. The phasing of this classroom work through teacher modelling and demonstration, to guided and collaborative small group and individual activity, and then to independent critical response, is dealt with in more detail in Chapter 7. This chapter focuses on describing practical approaches to analysing the different aspects of the texts that will inform the design of classroom learning activities.

The re-presentation of these stories in electronic media entails adaptation of the original to a greater or lesser extent. In some cases the text of the story remains unchanged. This is so with classic stories now published on the web such as Oscar Wilde's *The Selfish Giant* (http://www.kiddyhouse.com/Kids), and it is also the case with some relatively recent picture books now on CD-ROM such as *George Shrinks* (Joyce, c.1994) and *The Paper Bag Princess* (Munsch, 1994). In other stories, such as *The Polar Express* (Houghton Mifflin, 1995) and *The Little Prince* (de Saint-Exupery, 2000a, b), there are only minor changes to the text in the electronic versions. On the other hand, there are quite significant differences in the text among various CD-ROM, web and book versions of traditional tales like *The Three Little Pigs* (http://www.shol.com/agita/pigs.htm) and *Mulan* (Disney, 1998) and in the CD-ROM and book versions of some contemporary stories like *Stellaluna* (Cannon, 1996). The images in book and electronic versions most frequently do differ significantly. There are only a few exceptions to this among contemporary stories such as *The Paper Bag Princess* (Munsch, 1994) and, of course, scanned versions of picture books more than 50 years old and now out of copyright, which are available through online digital libraries. The use of animation and hyperlinks to create 'hot spots' for readers to click on seems to be largely confined to the re-presentation of stories on CD-ROM rather than on the web. The first section of this chapter compares CD-ROM stories with their original book versions. The second section deals with book and web versions of traditional and classic tales, focusing on a comparison of scanned older versions of classics in online digital libraries with more contemporary retellings in book format.

From book to CD-ROM: reading the multimedia reconstruction of narrative

We will begin with a discussion of variation in the images in the CD-ROM and book form of *George Shrinks* (Joyce, c.1994), which maintains the same text in both versions and is suitable for children in the early years of school. Maintaining the focus on images, we will then look at *The Little Prince* (de Saint-Exupery, 2000a, b), suitable for readers from pre-teen to adulthood. Next we will look at CD-ROMs where both the images and the text differ significantly from the book versions, such as *Stellaluna* (Cannon, 1996), another story suitable for younger readers, and also *Mulan* (Disney, 1998), which is most commonly used with children from about 8 to 10 years of age. For a further example of this kind of comparative analysis using texts suitable for senior high school students see Jewitt's (2002) comparison of the book and multimedia CD-ROM version of John Steinbeck's, *Of Mice and Men* (Steinbeck, 1937).

George Shrinks

This popular story by William Joyce (1985; c.1994) now forms the basis for an animated television series with associated website activities for children (http://pbskids.org/georgeshrinks/), but the focus here is on the original book and CD-ROM. In this story, George dreams, in his parents' absence, that he is small, and he awakens to find himself about the size of a mouse. Nevertheless, he undertakes the list of tasks his parents have left for him in a note – from brushing his teeth to looking after his baby brother. But these everyday routines turn out to be highly adventurous for the miniature George.

The CD-ROM version of *George Shrinks* has the same text as the book, but five images from the book have been deleted from the CD-ROM. The first of these is the second image of George making his bed. The next is the immediately subsequent image in the book of the miniature George standing next to the skirting board with the cat's back legs visible as s/he walks past (this image has no accompanying text). The third deleted image is of George tidying up his toy soldiers. The fourth deleted image is of miniature George on his brother's head frightening the cat. This image accompanies the text '... and play quietly'. The final omitted image has no accompanying text. It shows George flying his plane over the roof of the house. In all cases where images with text have been deleted the text has been added to that accompanying the previous image.

The other images common to the book and the CD-ROM are substantially the same, except, of course, for the animations on the CD-ROM. Some of the animations are automatically included in the CD-ROM presentation while others are controlled by the reader's use of the mouse to click on 'hot spots' or hyperlinks that activate the animation. The mouse-activated animations include some that are peripheral or irrelevant to the storyline while others are quite integral to the story. Some of the animations that involve the cat are presented automatically and some are mouse-activated. Both types of animations involving the cat are integral to the story, and it is the variation in the type and frequency of the images of the cat across the book and the two versions of the story on the CD-ROM that construct the different interpretive possibilities in these three versions.

In the book the double page spread immediately following the title page has copyright and other publication information superimposed on an image of miniature George on the right-hand page, hiding against the corner of the skirting board, holding a crayon and looking anxiously at the leg of the cat on the left-hand page. This image clearly foreshadows the threat of the cat. It does not appear on the CD-ROM. However, in the book the cat does not appear again until image number five (referred to above as George standing next to the skirting board with the cat's hind legs visible). The next appearance of the cat in the book is in image thirteen (also referred to above as George on his brother's head and the cat in fright). In the book the next appearance of the cat is in image twenty where it attacks George in his aeroplane. In subsequent images the cat pursues the attack in George's bed but is foiled upon the return of George's parents and George's return to his normal size.

In the book then, notwithstanding the foreshadowing in the image prefacing the story, there is no overt threat from the cat until image twenty near the end of the story. In the CD-ROM version where the story is read with animations but no hot spots, there is in fact only one image prior to image twenty portraying the cat as a threat. This is in the picture where George is eating cake for breakfast and the cat's paw lands on the handle of the spoon George is sitting in, catapulting him off-screen. In the CD-ROM version with hot spot access however, the foreshadowing of the cat as a threat is much more frequent and much more explicit. In the second image for example, clicking on George's robe resting on the drawer beside the bed results in the cat's paw 'swiping' at the robe as George is about to put it on. In the third image as George is making his bed, the cat's paw periodically swipes menacingly at the clothes hanger. In the image where George is brushing his teeth, clicking on the bathroom tile reveals the menacing gaze of the cat. The same breakfast image where the cat's paw lands on the spoon occurs in this version and in the image where George is collecting the mail, clicking on the cat again activates a menacing swipe of its paw.

Both CD-ROM versions also imply a greater intimidation of the cat by George than is the case in the book. We have noted that the image in the book showing George on his brother's head and the cat in fright is not included in the CD-ROM. However the prior image of George riding on his brother's back is common and in both CD-ROM versions this includes an animation where George jumps onto his brother's head brandishing a spoon, disappears off-screen and then jumps back onto his brother's head. In the background a cat's scream is heard.

What we have then is ostensibly the same story, in three different versions, each of which afford different interpretations of the role of the cat in the unfolding events, and in each case these differences are essentially achieved via the variation in the frequency and type of images involving the cat.

The Little Prince

The Little Prince was first published in 1943 and has been republished many times, most recently by Penguin Books (2000) and by Harcourt Inc (2000) in a larger format. Also in 2000 the CD-ROM with hyperlinked animations of

Figure 4.1 High angle view: Looking down on the Little Prince

Figure 4.2 An eye-level view of the geographer

the original images was published by Tivola/Gallimard (2000). In this story the narrator is a lone aviator whose plane suddenly has mechanical difficulties over the Sahara Desert and he is forced to land. A young boy comes up to him from nowhere and asks him to draw a sheep. Gradually the pilot learns that the Little Prince is a visitor from space who lived on a small asteroid named B-612. He left his home to explore, and describes his journey from planet to planet, each tiny world populated by a single adult. As the Little Prince recounts these visits, the author pokes fun at a king, a businessman, a geographer, and a lamplighter, all of whom signify some futile aspect of adult existence. Eventually the Little Prince is carried by a flock of birds to Earth. Here the parable or fable like story of *The Little Prince* continues, addressing deep philosophical issues about love, relationships, the emptiness of a life without either, death, spirituality, capitalism, and, in general, the soulless existence of the adult world. There are 46 drawings in the book – some coloured and some black and white. They are not naturalistic but rather more schematic representations of the participants and circumstances rather than realistic images. The animations in the CD-ROM maintain these features of the original images.

The story has 27 chapters and a total length of over 90 pages. Here we will deal with one episode in chapter 15 only – the encounter of the Little Prince with the geographer on his planet (for a more extensive discussion see Unsworth (in press)). In this segment the geographer is keen to record what the Little Prince can tell about his planet. The Little Prince mentions that his planet has two miniature volcanoes and a single flower, but the geographer will not record the flower because it is 'ephemeral'. This involves an interesting discussion about what ephemeral means and arouses feelings of regret in the Little Prince at having left his flower all alone.

In the book there is only one image in chapter 15 – a drawing of the geographer. This shows an elderly, white-bearded man sitting behind a large desk, looking at a large open book and holding a magnifying glass in his left hand. The old man and his desk are located on the top of an arc, which represents his planet. It is a medium to long view since we can see the entire desk and the top half of the man behind it as well as the arc of the planet, so although not remote, the viewer is socially distanced from the geographer. The vertical angle is high so we look down on the geographer. Composition- ally the image takes up about one-third of the page and is located on either the first or second page of the chapter depending on the edition.

In the CD-ROM all of chapter 15 is available optionally as a hyperlinked animation. The chapter is overwhelmingly dialogic and hence the spoken text of the animation is almost identical with the written text of the book. In the book there is no representation through images of the discussion about the ephemeral nature of the flower. In fact, there is only the one initial image of the geographer and none of the Little Prince in this episode. In the CD-ROM the discussion of 'ephemeral' is preceded by a discussion of the geographer's role. In this discussion the Little Prince and the geographer are both depicted together on screen with successive scenes, each depicting both characters and taking up only a section of the screen, alternating from the top left to the bottom right. The discussion of ephemeral is distinguished by the full screen presentation of the images. In this discussion the Little Prince and the geographer are presented separately and alternately as full screen views in scenes six to ten as summarized in Table 4.1.

In the discussion of the meaning of 'ephemeral' not only do all of the images occupy the full screen, but also they are very different from the images in the previous discussion in the ways they construct interaction with the reader/viewer. These images are much closer views. In each case the image is a demand with the character looking directly at the reader/viewer; all of the images of the Little Prince are high angle views, and all of the images of the geographer are at eye level; all of the images of both characters have a parallel horizontal angle, being 'front on' to the reader/viewer. These features are summarized in Table 4.2

In scenes six, eight and ten, the Little Prince is looking at the viewer, demanding our interaction. So, in these images we are looking at him from the point of view of the geographer looking over the desk, and the enormous geography book. These images are also elevated, so from the geographer's viewpoint, we are looking down on the Little Prince from a position of relative power.

In scenes seven and nine, however, the geographer is looking at us and we are positioned as having the point of view of the Little Prince. Here it is interesting to note that the vertical angle is much closer to eye level, so that from the point of view of the Little Prince, the geographer is not accorded greater power.

In all of the images in scenes six to ten the social distance is medium, with the upper body and head of the characters visible. The frontal plane of the characters is also parallel with that of the viewer indicating maximal inclusion.

Table 4.1 Animation scenes in chapter 15 of the CD-ROM version of *The Little Prince*

Scene	Screen layout	Characters	Synopses of chapter segments for each scene
6	Full screen	Little Prince	The geographer prompts Little Prince to describe his planet. Little Prince does so and ends by indicating his planet has a single flower.
7	Full screen	Geographer	The geographer indicates that geopgraphers do not take note of flowers despite their beauty because flowers are ephemeral.
8	Full screen	Little Prince	Little Prince asks what 'ephemeral' means. The geographer indicates that eternal things like mountains and oceans rarely change. The Little Prince notes that volcanoes come back to life and asks again what 'ephemeral' means.
9	Full screen	Geographer	The geographer replies that what is important to him is the mountain that does not change rather than its being a volcano that might. The Little Prince again asks what 'ephemeral' means and the geographer replies 'which is threatened by impending death.
10	Full screen	Little Prince	Little Prince notes that his flower is threatened by impending death, which the geographer confirms. Little Prince regrets leaving her alone with only four thorns to defend herself.

Table 4.2 Interactive features of images in the chapter 15 animation of *The Little Prince*

Scene	Screen layout	Character(s) on screen	Social distance	Offer/ demand	Vertical angle	Horizontal angle
6	Full screen	Little Prince	Medium view	Demand	Elevated	Parallel
7	Full screen	Geographer	Medium view	Demand	Eye level	Parallel
8	Full screen	Little Prince	Medium view	Demand	Elevated	Parallel
9	Full screen	Geographer	Medium view	Demand	Eye level	Parallel
10	Full screen	Geographer	Medium view	Demand	Elevated	Parallel

So in the discussion of 'ephemeral' the viewer is positioned to engage with the represented participants much more, and more intimately, and also from changing points of view than is the case in the previous segment concerning the geographer's role.

With the exception of scene eight, the character who is the principal speaker in the scene is the character depicted on the screen and is seen from the point of view of his interlocutor. This can be seen in Table 4.3, which shows the speakers and the number of words spoken in each scene with the principal speaker indicated in bold font.

Scene eight is the only occasion when the listener, who happens to be the Little Prince in this role, is depicted on screen. How is he depicted during the geographer's speech? The geographer is expounding on the virtues of geography books never going out of date because they record eternal things like the position of mountains. The Little Prince is seen from a high angle respectfully fingering the edge of the geography book on the geographer's desk and looking up at the viewer who is positioned as the geographer. So the visual construction of this encounter accords somewhat more humility to the Little Prince than is the case in the book and linear text version.

The analyses of the images in comparing the book and CD-ROM versions of this segment of the episode with the geographer helps students to understand what is meant by the textual construction of point of view and how this is achieved through images. In this case it also alerts students to the complex shifting of point of view in the CD-ROM version and how this differs from the linear text versions.

Stellaluna

In this story a baby bat, Stellaluna, is separated from its mother when she was avoiding an attack by an owl. Stellaluna lives in a nest with a family of young birds and adopts bird-like behaviours. Eventually Stellaluna and her mother are reunited but Stellaluna visits the birds she has made friends with and lived harmoniously, with despite their differences.

For the most part the main images in the book and the CD-ROM versions of this story are common, as was the case with *George Shrinks*. In *George Shrinks* some of the hyperlinked animations, such as those involving the cat, are strongly related to the story line, and some are either peripheral or irrelevant to the main story, providing playful distractions. In *Stellaluna* the hyperlinked animated images are predominantly superfluous to the story

Table 4..3 Visual point of view and principal speakers in the scenes of the 'ephemeral' discussion

Scene	Screen layout	Character(s) on screen	Social distance	Offer/ demand	Vertical angle	Horizontal angle	Who speaks	Point of view
6	Full screen	Little Prince	Medium view	Demand	Elevated	Parallel	**Little Prince 34** Geographer	Geographer
7	Full screen	Geographer	Medium view	Demand	Eye level	Parallel	Little Prince 7 Geographer 11	Prince
8	Full screen	Little Prince	Medium view	Demand	Elevated	Parallel	Little Prince 16 **Geographer 43**	Geographer
9	Full screen	Geographer	Medium view	Demand	Eye level	Parallel	Little Prince 5 **Geographer 29**	
10	Full screen	Geographer	Medium view	Demand	Elevated	Parallel	**Little Prince 33** Geographer 1	Geographer

and do not appear in the book version. For example, a number of activities of jungle animals are included which are quite unrelated to each other or to the story – such as a monkey running up a tree, elephants splashing water at each other, a giraffe drinking and then gargling, and a bird sliding down the giraffe's neck. The variations in the main images across story versions are more subtle, but they have quite a significant impact on the way some aspects of the story are interpreted differently in the different versions. Here we will look at just two examples – the initial attack by the owl and the first scene reuniting Stellaluna with the bats.

In the book the owl is presented only once in a distant view showing it approaching the mother bat and Stellaluna plummeting to the ground. The visual presentation of the owl attack is minimized. In the CD-ROM animation there are several scenes showing the owl attack, which reflects the text – 'the owl struck again and again'. These images follow a medium-close up view of the stationery owl with a menacing look directed towards the unsuspecting bats, again reflecting the text – 'as mother bat followed the heavy scent of ripe fruit, an owl spied her'. In the CD-ROM the owl is visually demonized and the owl attack is visually prominent, which is not the case in the book.

The re-uniting of Stellaluna with the bats is also represented differently in the images in the book and the CD-ROM. In the CD-ROM the scene where she meets the older bat who queries her unusual 'bird-influenced' behaviours is a close up 'demand' view of the old bat looking directly at the reader and demanding interaction. This clearly constructs the reader as having Stellaluna's point of view. In the book, this scene is depicted as an offer where neither character looks directly at the reader. The gaze of the older bat is directed at Stellaluna and the reader looks on in more interpersonal detachment.

A lot of additional text is included in the CD-ROM. For example, 21 additional lines of text are directly attributable to Stellaluna, whereas in the book only 15 lines are spoken or thought by Stellaluna. The additional text seems to be designed to make meanings explicit and remove the opportunity for inference, as in the additions 'I love you, Stellaluna' and 'I love you' (from Stellaluna to Mother Bat). The additions are also more colloquial ('I gotta eat something' and 'I wanna take a nap you guys') than the standard forms of English in the book. The effect of this is emphasized by the somewhat unfamiliar standard form of the text from the book that is omitted in the CD-ROM ('You slept at night?' gasped another. 'How very strange,' they all murmured.)

The changes in the language, different visual perspectives in the images, and distractive animations, significantly change the somewhat serious tone of the *Stellaluna* book to almost a slapstick approach to humour in the CD-ROM. In learning about these differences and how they are constructed through semiotic choices within language and image, young children can be introduced to the kind of metaknowledge that will facilitate their development as critical comprehenders and composers of multimodal texts.

Mulan

The ancient Chinese story of Mulan has been published in many different versions in books (Jiang, 1997; Lee, 1995; Nan Zhang, 1998; San Souci, 2000), film (Wizard Animated Classics, n.d.) and on CD-ROM (Disney, 1998). In ancient China Mulan learns that her father has been drafted into the Emperor's army. He is too old and ill to go to war, and since there were no sons in the family, Mulan disguises herself as a man and heeds the call to arms in her father's stead. She travels far away and serves as an outstanding warrior for the Emperor for many years. Eventually her courage brings her to the attention of the Emperor who offers her whatever she wishes as a reward. She simply asks to go home, where she surprises her family and most of all her comrades in arms who still thought she was a man.

The Disney CD-ROM, *Mulan Animated Storybook* (Disney, 1998), is designed as a game in which the reader seeks clues to access via hyperlinks the lost scrolls that tell the Mulan story. The dragon Mushu is Mulan's friend who provides the reader with game navigation clues and also functions as a participant in the story. The scrolls correspond to the five main screens in the story and each one outlines one segment of the tale. The events outlined in the scrolls are also constructed interactively by clicking on hot spots. For example in the Imperial City, the final story screen, Mulan defeats the enemy leader and rescues the Emperor. The animation of these events shows Mulan in battle with the enemy leader and also requires the viewer to assist by selecting melons and 'throwing' them at the guards and providing other kinds of assistance to Mulan and Mushu. The resolution is the rescue of the Emperor. In the Disney version Mulan is revealed as a woman in the battle at Tung Shao pass and the soldiers leave her behind when they go to the Imperial City. Mulan discovers the enemy leader has survived the battle and she goes to the Imperial City to protect the Emperor from him, however she appears on this mission wearing a long dress, clearly signifying female apparel.

A number of games are embedded in the main 'story-game'. Most of these are relevant to the story, such as clicking on the wall tiles in Mulan's house to find her father's armour. Some of the clicking activities, as in other CD-ROMs, are quite peripheral to the story, such as characters throwing snowballs at each other on the Tung Shao pass. Characteristic of the premium put on the bodily interactivity of the reader in the story is the fact that the reader needs to click on hot spots at crucial times, as a participant in the action, in order to advance the story. For example, when Mulan is rescuing the Emperor, Mushu instructs the reader to open the door (so that he can obtain more weapons). This direct reader involvement is essential to complete the rescue.

Unlike the other stories we have discussed, the Disney CD-ROM is not derived from any particular published book of the Mulan legend. This means there is little likelihood of commonality between the CD-ROM images and those in the book versions. Nevertheless, it is useful to consider the kinds of observations students might make by comparing images in book versions that present the story more traditionally with those on the Disney CD-ROM. In one traditional book, *The Song of Mulan* (Lee, 1995), the story begins with a medium to close up view of Mulan at her loom. Her departure for the army shows a slightly low angle view of Mulan on horseback looking down at her parents. Then the images are more distant, remote, high angle views of battle. After Mulan returns home and prepares to reveal her secret to her comrades in arms we see a close to medium view of Mulan's head and shoulders from the rear as she looks into the mirror, where we also see her reflection. Here it is as if we are positioned along with Mulan and aligned with her perspective. This kind of variation does not occur on the CD-ROM. All of the images are fairly distant views showing full-length images of the participants in the middle ground of the various scenes. The only exception is the introductory screen, which is a high angle view looking down on Mushu as he is entrusted with the scrolls of the Mulan legend.

The key elements of the plot of the story remain fairly consistent across a range of published retellings with the Disney CD-ROM representing perhaps the greatest departure, especially in the conclusion. The narrative technique is most commonly a third person narration with dialogue included. This is the case, for example, in *The Ballad of Mulan* (Nan Zhang, 1998). The scrolls in the Disney CD-ROM are also third person narration with dialogue occurring in the animations. There is a certain amount of redundancy between the narration and the dialogue, but the latter also provides some elaboration.

This can be seen in the early part of the story in Mulan's house where the scroll reads:

> Mulan's father had been called to serve in the army in the war against the Huns. She feared for his life and decided to go in his place.

In the animation we then hear directly from Mulan:

> By order of the Emperor one from every family must serve in the imperial army. My father is not well enough to join the army. I must go in his place.

In some cases the dialogue is a very colloquial form of English reflecting social practices of some particular groups in Western culture. This can be seen where Mulan is in the army camp and anxious to establish her male persona:

> I like to do manly things. Who's ready for a bar-b-que and burping contest?

On the other hand, *The Song of Mulan*, which is, quite unusually, told principally as a first person narrative, includes some elements that strongly invoke traditional Chinese culture. On the penultimate double page spread for example, there is an image of an old Chinese gentleman above the following proverb:

> A male rabbit is fast and agile,
> A female rabbit has bright eyes.
> When two rabbits run together,
> No one can tell which is male, which is female.

The emphasis on action in the Disney CD-ROM, compared with a strong representation of reflection in *The Song of Mulan*, can be explored initially by comparing the verb types in both texts. In the CD-ROM narrative for example, the verbs are mostly action verbs with very few verbs of thinking or perception. In *The Song of Mulan* however, there is a pattern of frequent use of the verb 'hear', indicating Mulan's reflection on her leaving home. On her first night away Mulan says:

I hear my father's voice no more,
Only the rush of the river.

Later she says:

I hear my mother's voice no more,
Only the enemy's horses neighing.

This drawing attention to what can be heard in the context of reflective thought, intensifies the effect of the narrator's voice on the following page:

'Ten thousand miles to the border.
I cross high mountains as if on wings.'
War drums ring in the brittle air.

Then, as Mulan returns home the use of 'hear' contrasts the joy of return with the earlier sadness of not hearing her family:

Father! Mother! Hear your daughter return.
...
Elder Sister! Hear Younger Sister return.
...
Younger Brother! Hear Second Sister return.

The Mulan legend has enduring appeal and its multiple versions provide different kinds of attraction and engagement as well as constructing different ideological perspectives through different narrative techniques and uses of language and image. Through comparative work children can enjoy the multiple story versions and the differences in the ways they are told as well as developing a critical understanding of how the linguistic, visual and digital resources construct these variations.

Back to the future: reading classic stories online

We have noted that while there has been some significant publication of contemporary children's literature on CD-ROM, very few contemporary books have been re-published on the web. On the other hand, there are now very extensive online collections of classic children's (and adult) stories. Some of these are text only versions, but there are also a large number of illustrated versions online. In this section then we will focus on comparing online illustrated classics with more recently published book versions. It is possible to undertake this kind of comparative work with students of all ages in the school system from the early years to older adolescents in the final years of school. Online resources provide access to classic books with limited amounts of text for early readers such as Edward Lear's *The Owl and the Pussycat* (http://www. storybookonline.net/article.aspx?Article=Owl_And_The_Pussy-Cat); and several versions of traditional tales such as *The Three Little Pigs* (http://www.shol.com/agita/pigs.htm; http://www.cupola.com/html/wordplay/3pigs.htm). For developing readers there are stories with somewhat more text such as Oscar Wilde's *The Selfish Giant* (http://www.geocities.com/Heartland/7134/Christmas/chrselfgiant1.htm). For experienced readers, illustrated online versions of famous works of literature are also available. For example, the University of Virgina e-text library has works of classic writers like Dickens' *A Christmas Carol* (http://etext.lib.virginia.edu/toc/modeng/public/DicChri.html) and Arthur Conan Doyle's *Hound of the Baskervilles* (http://etext.lib.virginia.edu/toc/modeng/public/DoyHoun.html).

Due to the constraints of space here, we will focus in this section on one story suitable for use with pre-teenage children. This is the story of William Tell.

William Tell

The story of William Tell has been popular for a very long time and its continuing appeal is evident in the publication of new versions at least every 10 years since 1981 (Bawden, 1981; Buff, 2001; Early, 1991; Fisher, 1996; Small, 1991). The story of the legendary Swiss archer compelled to shoot an apple from his son's head by the tyrant Gessler is almost always retold in a serious tone, emphasizing the bravery of Tell's son and the quiet but resolute nature of Tell's planned revenge with a second arrow, had his first caused harm to his son. However, a much more humorous version of the story was published in 1904. The International Children's Digital Library (http://www.icdlbooks.org) has made it possible for readers today to access a scanned copy of this version entitled *William Tell Told Again* written by P. G. Wodehouse and illustrated by Philip Dadd. Comparison of this online text with more contemporary book versions provides an amusing context in which students can extend their experience of narrative through online resources.

The humour of *William Tell Told Again* is achieved through the images and the text. This provides an engaging focus for children's work in learning about the grammar of language and images, enabling them to use grammatical concepts to describe how language and image are constructed to achieve quite different retellings of the 'same' story. We will discuss two key episodes – when Gessler orders Tell to shoot the apple from his son's head, and when Tell responds to Gessler's enquiry as to the purpose of Tell's second arrow. First, we will compare the humorous images of Philip Dadd with those in a recently published book version. Then we will compare the language choices constructing these episodes in the different versions.

The humour is immediately obvious in the Philip Dadd images. Figure 4.3 shows plate 10 from the 1904 book. Here we see Gessler in his fine clothes, on his horse with an apple on the top of his bald head, hand on his hip, looking quizzically at Tell and delivering his command about shooting the apple from the head of William Tell's son. The corresponding image from the Margaret Early book, Figure 4.4, is much more sombre. Here we see Gessler standing beside Tell who has his arm on his son's shoulder. Gessler has one hand on his hip and the other arm outstretched holding his falcon and pointing away from Tell in the act of issuing his command while one of the soldiers is carrying the apple. There is a good deal of difference then, in the content or what is represented in the image – the ideational meanings. But the interactive meanings – the way the represented participants relate to us as viewers – are also very different. In the Philip Dadd image we are much closer to the participants, positioned behind Tell and his son and looking up with them to Gessler. Since we are looking up to Gessler we are in a position of diminished power similar to Tell. Furthermore, we are looking at Gessler from a similar perspective to that of Tell, so we are closely aligned with his point of view. On the other hand, in the Margaret Early picture we are much further away, looking down on the whole scene so we are not positioned subordinately to any of the represented participants, nor is our point of view aligned with anyone in the picture. Neither of the images is naturalistic (true to life as in a colour photograph), but the Philip Dadd picture is 'realistic', while the purpose of the Margaret Early images is to present the story as a kind of stylized tapestry, as emphasized in the patterns representing the water and the sky.

There are also significant differences in the layout or compositional meanings of the images. In the Philip Dadd image the participants take up most of the image space whereas in the Early picture the participants are

Figure 4.3 Plate 10 from Wodehouse (1904) *William Tell Told Again*

more dispersed and the background is given more space. Gessler is prominently located in the centre of the Dadd image and Tell's salience in the
foreground is emphasized by the greater colour saturation of his hair and
clothes. In the Early picture there is not the same degree of salience afforded
to the main characters.

Figure 4.4 Margaret Early's image of Gessler ordering Tell to shoot the arrow

Once this kind of comparative image analysis is modelled for students, it is then possible for them to undertake scaffolded application of these ideas to analyses of other images, which they could do collaboratively and with teacher guidance. These might focus on further images of the same event from different books or a comparison of images by Early and Dadd dealing with a further event in the story, such as the scene where Tell explains the second arrow he had at the ready, as shown in Figure 4.5.

The language constructing the scene when Gessler orders Tell to shoot the apple from his son's head is very different in the Early text and the Wodehouse text. The relevant segments from each of these texts are shown in Table 4.4.

The focus of our work in comparing the same story events from different story versions is to appreciate not only 'what' is different across versions but also 'how' the language has been used to construct these differences. Students

Figure 4.5 Comparing images depicting Tell's second arrow

Table 4.4 Gessler orders Tell to shoot the apple from his son's head

Early (1991)	Wodehouse (1904)
'If you are such a fine marksman, Tell,' replied Gessler, 'let me see you shoot an apple from your son's head. If you succeed, you shall both go free – but if you fail, the punishment is death.'	'Now, Tell, I have here an apple – a simple apple, not over-ripe. I should like to test that feat of yours. So take your bow – I see you have it in your hand – and get ready to shoot. I am going to put this apple on your son's head. He will be place a hundred yards away from you, and if you do not hit the apple with your first shot your life shall pay forfeit.'

will immediately notice that the Wodehouse version is longer and will readily identify the information in that version that is not included in the Early version. This is the detail about the ripeness of the apple, that Tell's son will be 100 yards from him and that Tell's first arrow must pierce the apple. If we continue our comparison of the content (or ideational meanings), students may suggest that the Wodehouse version has more action in it. This can be verified by looking at the proportion of action verbs (material processes) across the texts as indicated in Table 4.5.

If students do not make this suggestion, the verbs can be circled and then categorized and Table 4.5 constructed on the white/blackboard to cue students to think about this difference. But the effect is not just due to the greater proportion of action verbs. If we look at the participant roles associated with the action verbs we will see that in the Wodehouse segment, six of the seven action verbs have an explicit goal or object (italicized below):

> should like to test *that feat of yours.*
> take *your bow*
> am going to put *this apple*
> *He* will be placed
> do not hit *the apple*
> *your life* shall pay forfeit.

Table 4.5 Proportions of different verb types in two versions of *William Tell*

	Relational	Material	Mental	Verbal
Early	2	4	1	1
Wodehouse	2	7	1	0

And in the seventh case the goal is implicit:

get ready to shoot (*the arrow*).

So for most of the time someone is doing something to someone or something else. In the Margaret Early text only one of the four action verbs has a goal, so participants engage in events but these don't seem to impact on others and the effect is to confirm the somewhat lesser intensity of activity in this text. A further grammatical contribution to this effect is the 'downranking' of the clause describing the action Tell is compelled to perform:

'let me see [[you shoot an apple from your son's head]].'

The clause in double brackets functions as if it were a thing, representing the phenomenon to be seen, so it is a grammatical element within the whole clause whose main verb is 'see'. This clause is thus referred to as a downranked, rankshifted or embedded clause. This embedding tends to 'freeze' the action into a noun function and hence reduce the impression of action within the passage. Students may compare this clause with those representing this aspect of the story in the Wodehouse text. Students could also be guided to re-write the Early clause without using the embedded clause:

'I w—pl— an apple on your son's head, and, if you are so sk—— with a bow, you w— easily h—the apple with your first arrow from 100 yards.'

or

'I want to see y— sk—w— y— b— , so I w—pl— an apple on your son's head and you m—- h— the apple with your first arrow from 100 yards.'

The latter sentence enables students to see how the phenomenon of the verb of perception 'see' needs to be a noun or function as if it were a noun: 'your skill [with a bow]'.

Students may also note or be guided to see the greater immediacy of the Wodehouse segment. In this text we are given the impression that things are about to happen. We have the modal verb complex 'should like to test' and the complex verbal group 'get ready to shoot', then the future continuous 'am going to put' and the future 'will be placed'. By contrast the Early text uses predominantly single present tense verbs: 'see', 'shoot', 'succeed' and 'fail'.

From the perspective of interpersonal meanings the Wodehouse text has Gessler make two quite direct commands to Tell: '... take your bow' ... 'and get ready to shoot'. The only command in the Early text is more indirect in the 'Let' form 'Let me see you shoot an apple from your son's head'. Also from the interpersonal perspective the use of the vocative 'Tell' is interesting. In the Wodehouse text this is put in the initial position in the clause, so it is an interpersonal Theme forming a significant aspect of the orientation of the clause: 'Now, Tell, I have here an apple ...' By contrast in the Early text the vocative is placed at the end of the clause: '"If you are such a fine marksman, Tell," replied Gessler ...'

If we consider the topical Themes in both texts (the participant in the first position in the clause and hence the point of departure or orientation of each clause), we can see how much more the Wodehouse text constructs Gessler as the point of departure, as indicated in Table 4.6.

Table 4.6 Topical Themes in two versions of an episode in the William Tell story

Early (1991)	Wodehouse (1904)
If you	'Now, Tell, I
replied Gessler,	I
let me	So take
If you	I
you	and get ready to shoot
if you	I
the punishment	He
	and if you
	your life

In the Wodehouse text Gessler ('I') is in Theme position four times and in the Early text this occurs only once ('let me').

The corresponding passages in the Early and Wodehouse texts where Tell explains his second arrow are shown in Table 4.7.

Students will readily note the humorous approach taken in the Wodehouse text. This is most obvious when Tell is asked to explain 'one' thing and replies 'A thousand, your Excellency', to which Gessler responds 'No, only one'. But students may also note the repetition and techniques for delaying the 'punch line'. First we have the repetition of the question about the purpose of the arrow:

Table 4.7 Tell explains his second arrow

Early (1991)	Wodehouse (1904)
As Walter and his father were about to walk away, surrounded by townsfolk, **Gessler** called out: 'One moment, Tell. That was very clever, I admit, but before I let you go there is something you must explain. I am curious to know why you hid a second arrow in your jerkin …'	**Gessler** leaned forward in his saddle. 'Tell, he said, 'a word with you.' Tell came back. 'Your Excellency?'
	'Before you go I wish you to explain one thing.'
	'A thousand, your Excellency.'
	'No, only one. When you were getting ready to shoot at the apple you placed an
Tell gazed calmly and fearlessly at **Gessler**. 'Had my aim been false, and I had killed my son, that arrow would have been aimed at your heart – and I would not have missed,' he said.	arrow in the string and a second arrow in your belt.'
	'A second arrow!' Tell pretended to be very much astonished, but the pretence did not deceive the Governor.
	'Yes, a second arrow. Why was that? What did you intend to do with that arrow, Tell?' Tell looked down uneasily, and twisted his bow about in his hands.
	'My lord, he said at last, 'it is a bowman's custom. All archers place a second arrow in their belt.'
	'No, tell,' said **Gessler**. 'I cannot take that answer as the truth. I know there was some other meaning in what you did. Tell me the reason without concealment. Why was it? Your life is safe, whatever it was, so speak out. Why did you take out that second arrow?'
	Tell stopped fidgeting with his bow, and met the Governor's eye with a steady gaze.
	'Since you promise me my life, your Excellency,' he replied, drawing himself up, I will tell you.'
	He drew the arrow from his belt and held it up.
	The crowd pressed forward, hanging on his words.
	'Had my first arrow,' said Tell slowly, 'pierced my child and not the apple, this would have pierced you my lord. Had I missed with my first shot, be sure, my lord, that my second would have found its mark.'

'Yes, a second arrow. Why was that? What did you intend to do with that arrow, Tell?'

Further on in the passage we have effectively an additional three repetitions of the same question:

'Tell me the reason without concealment. Why was it? Your life is safe, whatever it was, so speak out. Why did you take out that second arrow?'

Then we have several techniques delaying and dramatizing the moment of Tell's response. One is an announcement that he will answer the question:

'Since you promise me my life, your Excellency,' he replied, drawing himself up, 'I will tell you.'

Then, when he does answer, this is interrupted by the 'included' projecting clause:

'Had my first arrow,' said Tell slowly, 'pierced my child and not the apple ...'

Students can be guided in understanding further how the language choices construct different interpretations in these accounts of the same incident in different versions of the story by exploring other aspects of the grammatical construction of interpersonal, ideational and textual meaning. From the perspective of interpersonal meaning they could note the distinctive use of the vocatives referring to Gessler in the Wodehouse text: 'Your Excellency', 'Excellency' and 'My lord'. Another difference is that in the Early version there are no direct questions. The form of the single question in this text is a clause projected by the mental process (thinking verb) 'know':

'I am curious to know why you hid a second arrow in your jerkin ...'

Whereas in the Wodehouse version there are number of direct questions:

'Why was that? What did you intend to do with that arrow, Tell?'
...
'Why was it?'
...
'Why did you take out that second arrow?'

The more dialogic nature of the Wodehouse text achieved by such resources increases the immediacy and dramatic effect and enhances the more humorous approach.

From an ideational perspective, at the beginning of the Wodehouse extract the grammar of Gessler's request to Tell indicates that Tell has secured his freedom:

'Before you go I wish you to explain one thing.'

On the other hand, in the Early text, Gessler is grammatically the Agent in the clause, indicating that Tell's freedom remains in Gessler's hands:

'... before I let you go there is something you must explain.'

Students can be guided to note the difference in the way Tell responded to Gessler about the second arrow. In the Early text qualities of calmness and bravery are ascribed quite directly and without equivocation to Tell:

Tell gazed calmly and fearlessly at Gessler.

In the Wodehouse text Tell is initially anxious, and only after he has received reassurance of his freedom does he present a bolder demeanour to Gessler. But his initial anxiety and subsequent boldness are invoked from the account of his actions rather than the direct ascription of the Early text:

Tell looked down uneasily, and twisted his bow about in his hands.
...
Tell stopped fidgeting with his bow, and met the Governor's eye with a steady gaze.

In the Early text then the opposition between Tell and Gessler is stark and direct, whereas in the Wodehouse text it is a more subtle duel – with some humorous overtones.

From the perspective of textual meaning, students may perceive that the Wodehouse text is more clearly oriented to Tell as the point of departure in the development of the account of the event. This is reflected in the number of times Tell is selected as Theme, located in first position in the clauses of the narrator:

Tell came back.

...

Tell pretended to be very much astonished ...

...

Tell looked down ...

...

Tell stopped fidgeting ...

...

he replied

...

He drew the arrow from his belt ...

By contrast in the Early text, Tell is in Theme position only once:

Tell gazed calmly and fearlessly at Gessler.

Also from the perspective of textual meaning, the effect of placing the dependent clause in Theme position (that is, before the main clause) may be noted. Students can observe this easily in both texts when the dependent clauses 'Before I let you go ...' and 'Before you go ...' are placed before Gessler's request for an explanation about the second arrow. Students should be made aware that it would be perfectly grammatically correct to re-order the clauses to place the dependent clause in the second position.

This understanding enables students to see the thematizing of the dependent clause as a key grammatical resource in these texts. They can appreciate the way in which the emphasis by Tell on the security of his freedom is achieved grammatically and how this resource also enables the cause to be placed before the effect in the grammatical construction of the account.

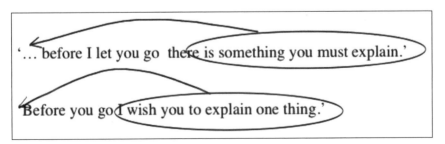

Figure 4.6 Dependent clause as Theme

Since you promise me my life ... I will tell you.

In both texts this resource enables the purpose of the second arrow to be located in the position for new information at the very end of the clause complex (when it would have been grammatically correct to place this final clause in first position), thus heightening the dramatic effect.

> 'Had my aim been false, and I had killed my son, that arrow would have been aimed at your heart ...'
>   'Had my first arrow,' said Tell slowly, 'pierced my child and not the apple, this would have pierced you, my lord. Had I missed with my first shot, be sure, my lord, that my second would have found its mark.'

In the Early text the effect is heightened by the inclusion of two dependent clauses and in the Wodehouse text by the repetition of the thematic dependent clause structure in the two sentences.

In this section we have shown how children can learn a great deal about the visual and verbal 'constructedness' of text using only two very brief story segments. There are many more opportunities for enjoyable learning afforded by both story versions. One such aspect of the Wodehouse text that should be mentioned is the inclusion of the captions that accompany each illustration. The following caption accompanies the image dealing with the second arrow:

> But, as the arrow cleft the core,
> Cried G. with indignation,
> 'What was the second arrow for?
> Come, no e-quiver-cation!
> You had a second in your fist.'
> Said Tell, the missile grippin',
> 'This shaft (had I that apple missed)
> Was meant for you my pippin!'

Conclusion

The publication of existing literary texts in multimedia electronic formats creates a different narrative experience of the original texts. Even scanned classic texts now provided online can be experienced in ways not possible with the original books – multiple pages can be viewed simultaneously on screen, images can be enlarged, these and text can be downloaded – and the online versions are readily available gratis, facilitating comparison with recently published book versions. This chapter has emphasized the construction of different interpretive possibilities of the 'same' story in book and electronic formats by comparing the nature and role of images and the variation in language across versions. But the interpretive possibilities within the electronic versions also vary according to the material interaction of the reader with the affordances of the electronic media. More specifically, in relation to the kinds of electronic texts discussed here, whether and when readers take up optional access to hyperlinks of various kinds. One example of this was noted in the discussion of *George Shrinks*. If one reads the CD-ROM version with hot spots available and actually accesses the relevant hot spots, one might experience the much stronger foreshadowing of the threat of the cat than is the case in the book version or in the electronic version without hot spots. But failing to access the hot spots means this increased foreshadowing of the cat threat would not occur. In *The Little Prince*, if one clicks on the geographer's book during the discussion of 'ephemeral' the word first appears in the book and is then erased, drawing visual attention to the essence of the discussion. However, if one does not click on the book or does not click on it within a certain period of time, the appearance and erasure of the word in the geographer's book will not occur. Similar effects of the actual version of the electronic text experienced by the reader and consequent influences on their story interpretation can occur with scanned versions of traditional and classic texts. In such a version of Dickens' *Christmas Carol* for example (http://etext.lib.virginia.edu/toc/modeng/public/DicChri.html), the images appearing initially on screen are in fact 'cropped' images. These are perfectly coherent and appear complete, but one needs to know that it is necessary to click on the image to reveal a more encompassing image. In this story for example, the image of Marley's ghost on page 22 initially does not show the ghost at all, but simply Scrooge sitting in his red chair and looking to his right. Clicking on Scrooge reveals the entire image with Marley's ghostly figure looking down at Scrooge. This feature of images is common in a number of such scanned classics in this collection. The interpretation of e-literature, then, also entails reading practices that take account of the semiotic affordances of the digital media. We will pursue this dimension of literary experience further in the next chapter as we explore the nature of new literary texts in electronic format only.

Chapter 5

Emerging digital narratives and hyperfiction for children and adolescents

Introduction

New electronic forms of literary narratives for children and adolescents are emerging on the web and, to a lesser extent, on CD-ROMs. This chapter discusses examples of these new digitally originated literary texts, proposing a simple framework for describing the range of narrative forms and their compositional features, and indicating ways in which classroom work with children can address these new forms of narrative experience. In this introductory section, five distinct types of digital narrative are identified as well as two types of digital poetry and two categories of digital comics. (The integrative electronic game/narrative is dealt with separately in the next chapter.) Five aspects of the compositional features of these texts are also identified and a tentative sketch of a framework relating the types of texts and their compositional features is provided. In the subsequent sections of the chapter, examples of each of the types of digital literature are discussed and practical teaching/learning activities are indicated.

The forms of digital narrative for children and adolescents seem to fall into the following categories:

- *e-stories for early readers* – these are texts which utilize audio combined with hyperlinks to support young children in learning to decode the printed text by providing models of oral reading of stories and frequently of the pronunciation of individual words;
- *linear e-narratives* – these are essentially the same kinds of story presentations which are found in books, frequently illustrated, but presented on a computer screen;
- *e-narratives and interactive story contexts* – the presentation of these stories is very similar to that of linear e-narratives, however the story context is often elaborated by access to separate information about

characters, story setting in the form of maps, and links to factual information and/or other stories. In some examples it is possible to access this kind of contextual information while reading the story;

- *hypertext narratives* – although frequently making use of a range of different types of hyperlinks, these stories are distinguished by their focus on text, to the almost entire exclusion of images;
- *hypermedia narratives* – these stories use a range of hyperlinks involving text and images, often in combination.

Digital poetry can be distinguished according to whether or not it invites physical interactivity by the reader. Some poems, referred to here as 'e-poetry', are dynamic, multimodal presentations but with no physical interaction by the reader, as opposed to digital poems, which do include this reader interactivity, referred to here as 'hyperpoetry'.

'E-comics' are distinguished here from 'animations'. The former are comic strips that appear on screen and are composed of essentially still images with speech balloons. Although some contain some minimal dynamic images, no use is made of hyperlinks. Animations are similar to animated movies seen in the cinema, on television or on video and/or dvd. The story animations that appear on the web frequently include some opportunities for viewer interaction, although these appear as adjunct activities interpolated into the story rather than an integral aspect of it. The focus here will be on e-comics.

E-fiction does not necessarily make use of the range of affordances of the digital medium. We can think of the 'take-up' of the various affordances as a cline. Some categories of digital fiction, for example, make no use of hyperlinks at all. They are at the extreme linear end of the cline from linear to hyperlinked stories. Some stories use hyperlinks between chapters or episodes only. These can be thought of as located about halfway along the cline. Other stories use hyperlinks to link episodes, and elements of stories within and across episodes and/or story versions. These might be thought of as at the extreme hyperlinked end of the cline. We can similarly think of clines of other dimensions of the compositional features of the stories. Monomodal texts use print only, whereas multimodal texts use print as well as all or some of the other available semiotic resources such as images, sound effects and music. Image use may be on a cline from still images only, to images with some dynamic features (such as a character's arm that waves), to fully animated images. Reader interactivity may be on a cline from very little, such as scrolling down the screen only, to highly interactive, such as responding to hyperlinks and navigating through optional story paths. Another cline relevant to these stories is the extent to which metafictive elements are introduced, that is, the extent to which in the telling of the story, its artifice as a story is specifically highlighted. The broad relationships of these compositional features to the five main types of digital fiction identified above are indicated in Figure 5.1.

The framework shown in Figure 5.1 indicates that e-stories for early readers, linear e-narratives and e-narratives with interactive story contexts are often fairly similar on two dimensions of digital rhetorics: they are linear stories with no metafictive elements. The e-stories for early readers do sometimes include dynamic images. These three story types vary most on multimodality and interactivity. The e-stories for early readers tend to be higher in multimodality since they often include music and sound effects.

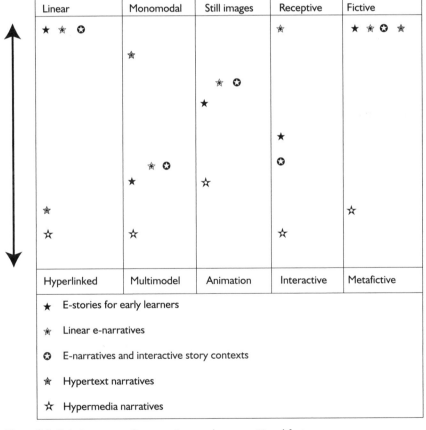

Figure 5.1 Relating types of e-narratives and compositional features

They afford significant interactivity, but only in the process of using the mouse to activate the reading of the text and of individual words. The e-narratives with interactive story context also potentially actively involve the reader in exploring the story background via hyperlinks. The linear e-narratives however, are receptive texts. They do not involve very much by way of physical interactivity by the reader, although they may be quite engaging stories. The hypertext narratives make use of varying kinds of hyperlinks to varying degrees and are usually quite high on interactivity, however they are frequently monomodal, containing no images and no metafictive elements. The most innovative of the categories of digital fiction are the hypermedia narratives (as well as e-poetry and hyperpoetry). No hierarchical valuing of different types of digital fiction is suggested here. Linear e-narratives can be just as engaging as hypermedia narratives. Developing students' appreciation of the nature of digital storying involves providing opportunities to make explicit to them how the affordances characteristic of particular types of digital fiction are deployed to achieve their narrative effects. In the following sections aspects of the form of the different types of digital fiction will be discussed and ways of alerting young readers to an understanding of their role in narrative will be explored.

E-stories for early readers

Stories for early readers composed for online distribution only, seem to focus mainly on supporting children in learning how to decode the text. These books frequently include dynamic images and sometimes animations. They are sometimes picture books but not storybooks, as is the case with *The Farm Animals* (Merino, 2004) on the 'Children's Storybooks Online' site (http://www. magickeys.com/books/). These e-stories (and picture books) have the facility for the text to be read to the young viewer, usually activated by clicking on a loudspeaker icon positioned in front of the text. In some of these stories it is also possible to click on each individual word in the text and have the word pronounced. However, in other stories the type of decoding support offered is more limited, amounting to the text being automatically read sentence by sentence upon clicking, with the whole of the target sentence being highlighted throughout its reading. There is no facility for individual words to be pronounced. In some cases the stories are somewhat mundane, however many others do make literary connections as is the case on the Tumblebooks site (http://www.tumblebooks.com/syndication/chickadee/

indexwf.html) with interesting versions of *Old Mother Hubbard* (Steen, 2000b) and *Jack and the Beanstalk* (Steen, 2000a). On the other hand, some of the e-books on sites such as Children's Storybooks Online (magickeys) are original stories with some literary quality in their own right. *Wolstencroft The Bear* (Lewis and Moore, 2003) is a story, richly illustrated with static images, of a bear who remains unchosen by buyers in a second hand shop until eventually a young boy with the same name comes into the shop with his father. As well as dealing with names that are burdensome for children, the story also deals with the departure of friends and loneliness as well with acceptance of oneself.

Wolstencroft ... provides an opportunity to develop young readers' understanding of the visual and verbal construction of point of view. Children will readily indicate their sympathetic feelings for Wolstencroft, sharing his loneliness, disappointment and his eventual happiness with his new family. It is possible to help young readers to become alert to some of the ways in which the authors and illustrators encourage this kind of alignment of the reader's response with the point of view of the focal character. One useful strategy is to enable the children to see that the author tells us a lot about what Wolstencroft was thinking and feeling as well as what he was doing, so since we know a lot about his thoughts and feelings we are likely to understand him and feel that we 'know' him. By teaching young children to distinguish among different types of verbs such as action verbs, being verbs, thinking, feeling and seeing verbs, they can begin to notice that the vast majority of verbs with Wolstencroft as the subject (doer, thinker, feeler or see-er) are in fact thinking and feeling verbs. This can be done online with children by selecting 'edit page' from the 'file' menu in Netscape and then working with them to underline all of the verbs associated with Wolstencroft on the second page of the story. Then the different kinds of verbs can be colour coded, for example, by changing the font for the action verbs to red, the thinking verbs to blue and the feeling verbs to green (Figure 5.2). The children will see that there are only three red action verbs involving Wolstencroft as doer, but a total of eight thinking and feeling verbs with him as subject. This helps the children to understand how the writer emphasizes Wolstencroft's inner self.

By similarly introducing young readers to aspects of visual grammar, it is possible for them to begin to understand how the creation of particular types of images can influence readers to be aligned with the point of view of particular characters. One approach with young children reading this story is to ask them to imagine that selected images are photographs that they had

He had arrived in the store just before Christmas when there had been a lovely big tree in the window, all decorated with fairy lights. Yards and yards of sparkling tinsel had been draped over everything, and holiday music had been playing all the time. Wolstencroft was especially fond of Jingle Bells. He liked its light, tinkling sounds. It always made him feel merry.

At that time there had been lots of other bears to keep him company. In fact, there had been so many teddy bears crowded onto that one narrow shelf that he had scarcely had room to move.

But, one by one they had all gone. Gleefully waving goodbye as they were carried off to their new homes. Until finally, he was the only teddy bear left in the entire store.

He had hoped that Santa Claus would drop by on Christmas Eve and deliver him to a good home. But he hadn't. Santa had been too busy that year, delivering even more presents than usual.

Wolstencroft felt sad and lonely sitting there all by himself on the shelf that was high above the Christmas cards. He longed to have a child take him home and love him and play with him. But, most of all, to hug him. For no hug is ever too big for a teddy bear.

He was trying hard not to cry because he knew that tears would make his eyes all puffy and red and then he would have even less chance of finding a home.

But why oh why didn't someone choose him?

Why, he wondered, was he passed over so many times for other less beautiful bears?

Figure 5.2 Wolstencroft mainly associated with 'thinking' and 'feeling' verbs

taken and then ask them to work out where they would be positioned as the photographer in relation to the represented image. This can be done as a drama activity by having some children pose representing the characters in the images and having other children use a digital camera to take the photographs. The image on page two (Figure 5.2) shows Wolstencroft on a shelf with other bears slightly above eye line and looking directly at the viewer. This 'demand' image makes direct interpersonal contact with the viewer, emphasizing a pseudo visual interaction with the represented character, so, as readers, we are strongly engaged with Wolstencroft, albeit at a distance. Then on page four the image shows the two rabbits leaving with their new owners and we have a rear view of the humans with the rabbits looking back from the bags they are being carried in. However, in the foreground we have a partial side view of Wolstencroft's face looking at the departing rabbits. It is a close-up view of his face and the vertical angle of the image is at eyeline for Wolstencroft's face but looking up at the rear view of the departing humans (Figure 5.3). So, if we were photographing this image we would need to be standing very close indeed to Wolstencroft, at the same level as him, looking

Figure 5.3 Reader alignment with Wolstencroft's point of view

mainly at the departing rabbits and their owner and peripherally at the side of Wolstencroft's face. This image then visually aligns us with Wolstencroft's point of view.

These are very concrete ways in which young readers can begin to learn about narrative technique while simultaneously engaging with and enjoying the story. Other narrative techniques such as foreshadowing could also be taught through very concrete examples, such as the salience of Wolstencroft's name in the language and the image of the first screen in the story. This e-story could be used in the classroom in conjunction with many other well known 'teddy bear' stories in conventional book format dealing with similar themes, such as *Corduroy* (Freeman, 1975) or the more challenging recent picture book also set around two bears in a second hand shop, *Hyram and B.* (Caswell and Ottley, 2003)

Linear e-narratives

A significant proportion of original contemporary online e-narratives for children and young people make no use of hypertext as a narrative device. These are stories which one reads on screen, scrolling down to read the print in much the same way as one would by turning the pages of a book. In some cases there is a table of contents and one can click on the chapter heading to navigate to that part of the book. At the end of each chapter there is typically an arrow (or similar icon) that you can click on to advance to the next chapter or return to the previous chapter (or to the table of contents). Some stories in this format have no illustrations and resemble printed short stories or novels presented on a computer screen. Examples of stories of this kind can be found on the Sundog Stories website (http://www.sundogstories.net/) – *St Mary's Avenue* (Ponzio, n.d.) explores Italian racism, aging and community, and the personal costs of militarism and war; *Flying Angels* (Ponzio and Labate, n.d.) takes place in late summer of 1850 when the children of a farm family befriend an itinerant bookseller who is accused of murdering the local storekeeper. Some of these linear e-narratives do include illustrations, but these are almost invariably static images. The Children's Storybooks Online website (http://www.magickeys.com/books/index.html) in the categories of stories for older children and for young adults, includes such illustrated linear stories.

Since these linear e-narratives do not make use of hyperlinks, audio or music, it might be expected that the electronic medium would be exploited to include some distinctive deployment of various types of images. However, the ways in which images are used seems to relate much more to the authors' style than to the affordances of the electronic medium. For example, on the Children's Storybooks Online site two very different stories by the same author make consistent use of fairly restricted categories of images. *Wind Song* (Moore, 2001) tells the story of an American pioneer farming family whose daughter's pet yellow bird saves them from a tornado. This story includes five images and a map. All of the images are long distance views, all are presented from an oblique horizontal angle, and all are either eye level or high angle vertical views. These choices construct visual remoteness between the represented participants and the viewer. There are no close-up images and no images where the characters are looking at the reader (demands). The same kind of visual remoteness dominates *The Littlest Knight* (Moore, 1994). In this story the king has indicated that suitors for the princess must rid the kingdom of the dragon that had been terrorizing the country. The Littlest Knight, son of a blacksmith, in his makeshift armour, eventually, through an act of kindness, breaks the spell that made the dragon behave antagonistically and secures him as a faithful guardian of the kingdom. There are 17 images in this story and all but two of them are distant views, usually high vertical angle and oblique horizontal angle views, again constructing remoteness between the represented participants and the reader. This remoteness is apparent in Figure 5.4 where the Littlest Knight is indicating to the King that he can rid the kingdom of the menace of the dragon.

Because there are only two images that are significantly distinctively different from that shown in Figure 5.4, these are important to note. The first of these is in fact the second image in the story (Figure 5.5), which shows the rear view of the Littlest Knight looking out of his window at the castle in the distance and the seemingly unattainable princess.

This close-up, eye-level view, where horizontal angle is parallel with the plane of the Littlest Knight, positions the reader to see from the same point of view as the hero of the story, thereby visually working to align the reader with his point of view. The last image in the book is also distinctive in that it is a demand showing the Littlest Knight and his princess side by side looking straight at the reader – engaging the reader interpersonally (see Figure 2.4).

These kinds of observations about images might provoke discussion with students about the impact on the narrative of making other kinds of choices about the types of images used to illustrate various aspects of the stories. For example, more close-up demand images or more aligning of the reader with

Figure 5.4 Distant view with high vertical angle, oblique horizontal angle constructing remoteness from the reader

particular characters' points of view, might have influenced quite significantly the interpretive possibilities of the text. Students can be helped to adopt this kind of critical visual reading by contrasting observations such as the above with similar stories where the image choices were different. For example, on the Children's Storybooks Online site, the story entitled *Second Thoughts*

Figure 5.5 Reader alignment with the point of view of the Littlest Knight

(Moore and Paulhamus, 2003) makes quite strategic use of demand images. In this story the Xxlepis from outer space visit the USA, bringing a powerful bacterial gift that greatly enhances the life of humankind. However, the President of the USA, and human society more generally, cannot understand that material value is not the kind of acknowledgement that the Xxlepis appreciate. As Moomba, a kind of ambassador for the Xxlepis, is advising the President about their values, he is looking directly at the reader. This is a demand image, engaging the reader optimally and positioning the reader with the President to receive the lessons in the account of the Xxlepis' value system. Once these kinds of observations and comparisons are modelled for students, they can begin to critically explore independently and in collaborative groups the choices of images in other linear e-narratives.

E-narratives and interactive story contexts

While most e-narratives are published online as 'stand-alone' stories, some are presented along with a range of additional story-specific information and related activities both online and offline, which extend readers' involvement with the particular story world. Some story sites include pages with additional information about characters, news of sequels or extensions to the stories, background information about the story world, hard copy companion publications or downloads, author biographies and email feedback on the story, story games and activity pages, quizzes and contests, newsletters, screensavers, and various kinds of merchandising such as the sale of t-shirts with story logos and captions. These reader-interactive story contexts create a distinct category of otherwise quite different e-narratives. Three examples are discussed here. The first is the story of *Banpf* (http://www.banph.com/) (Left Handed Creations, 1994–2004), the second is *Dead of Night*, an online story from The Nightmare Room (http://www.thenightmareroom.com/) of R. L. Stine, author of the *Goosebumps* stories, and the third is *The Inner Circle* (Matus, 2002), which is the first online book of *The Relic Triangle Trilogy* (http://www.relictriangle.com/). It is possible, of course, to generate learning experiences for young readers based on the use of language and image in the construction of these narratives, but due to constraints of space here, the focus of the suggestions for teaching will be on using the interactive story contexts provided online.

Banpf

The story of *Banph* occurs in a future world where humans, and mammals in general, no longer exist, and insects rule the earth. Many insect species have evolved and adopted human characteristics, leading an existence which parallels that of the human Middle Ages. Banpf is reputed to be the unluckiest of ants. Born without wings he has lost any claim to the title of prince, and the pampered life or privilege his fellow carpenter males enjoy. He has carved a place for himself as commander of his queen's army. He is more rugged, resourceful, and disciplined than his peers and has a wisdom beyond his years. Banpf endeavours to defend his kingdom against enemies of the carpenter empire, and encounters such foes as red slavers, barbaric beetles, and marauding locust.

The story extends over 12 chapters of about 3,000 words each and is richly illustrated with about six coloured static images interpolated throughout each chapter. The additional contextual information provided with this story includes:

- a brief history of the story's composition;
- a file of character sketches showing images of each character and a paragraph outlining, to a greater or lesser extent, each character's main personality traits, key behaviours and something of their role in the narrative;
- a similar file of sketches of future characters to appear in later Banpf stories;
- an outline guide to the nature of the future world in which the story is set;
- an email facility to send email to Banpf;
- examples of letters from reader correspondents;
- and a planned 'Kids Page' and 'Scrapbook' (under development).

The webpages of character sketches of the current and proposed future characters in the story can be explored prior to, during and/or after reading the story. They can also be modified to enhance interactive reading possibilities. One way to do this is to use the edit page facility in Netscape to add, change and or include examples of the personality features and behaviours of the characters, while progressing through reading the story. The edited page can then be saved into a folder on the computer and re-opened for further editing during subsequent reading sessions. Students may choose to collaborate on such tasks sharing the saved files. They may choose to make separate entries, colour coded by author. These augmentations might also include additional images in support of changed or additional descriptions of character traits. Such images could be saved from their locations in the main story and then inserted into the modified character description at the appropriate point. A further elaboration may be cropping (or otherwise modifying the image) with graphics software such as Graphic Converter prior to re-deploying the image in the character sketch. Students or collaborative groups who have produced such revised character files may then share their files and compare the modifications/elaborations they have made and discuss the justification for the various changes. Similar procedures could be used to modify the existing chapters of *Banpf* during or after reading the story as whole, by interpolating images of the future characters and editing the existing chapters to provide a role for them in the current story, consistent with their character description. Again, students could work individually or in groups on such tasks and then share the result of their work.

Dead of Night

The *Dead of Night* story is located in R. L. Stine's Nightmare Room (http://www.thenightmareroom.com). It is a 13-chapter characteristic horror story for young readers. Each chapter is accessed by clicking on the chapter number located at the bottom of the story window. The text of the story appears in a window internal to the story window surrounded by darkness, a spotlight, and indistinct partial image of someone's face and the title of the story. The contextual information on the other pages/windows in the site does not relate to this particular story but rather to other R. L. Stine stories. A second free online story is provided – *Your Own Personal Nightmare*. This has a similar format to *Dead of Night*, but is in the 'choose your own adventure' genre with hyperlinked choices at the end of some chapters. Another window provides access to sample readings of short segments of other R. L. Stine hard copy books. There is a window indicating that some titles will soon appear as e-books, an R. L. Stine screensaver and a link to the Warner Brothers Kids website (http://kidswb.warnerbros.com/web/home/home.jsp). There is also a '.pdf' file of the R. L. Stine writing program for children and printable Nightmare Room writing paper.

The Relic Triangle

This quest story is set in an ancient mythical land. In the first online book, *The Inner Circle*, young Kreelos Urandees has become heir to his father's kingdom, which has been wracked by civil war. One rising force, the Inner Circle, possesses The Crystal Vase of Purity, a relic of immense magic able to harness the powers of nature. The Inner Circle threatens to take over the kingdom of Atracia, now ruled by Kreelos. Kreelos must unite the disparate forces of his kingdom to overcome the threat.

The contextual pages included on the story site fall into five categories of interactivity in relation to reading the story. The first category, represented by the Overview link, might be considered 'anticipatory' or 'preparatory'. It provides a compendium to the Relic Triangle Trilogy, summarizing the online and hardcopy follow-up stories. The other categories of linked webpages are shown in Table 5.1.

The interactivity of the maps and glossary are intrinsic to the story. The online map of Atracia allows the reader to track the locations of the episodes in the story. It has navigational arrows to move around the map and this, with the capacity for zoom in and out so that locations of interest can be magnified by factors of five or ten, allows the reader to pin-point detailed locations. It is possible to download a free .pdf version of the online story as plain text. One possible learning activity using the online maps is for readers to save as image files the magnified locations they choose to examine using the online map and insert these into the plain text file of the story they have downloaded. Students could compare their 'map tracking' with that of other readers and discuss the significance of the sites they have chosen.

The glossary is a listing of all the names, places, and terms that are uncommon to newcomers to *The Relic Triangle Interactive*, as indicated in the following entry for 'barpin':

> barpin (bär' pin) n. – A unit of money. The basic Eldanese standard for one unit of money. 10 barpins equals one garlot. Yuladium coins are solely minted in the Grand Barpin House in Omifica (of Taran) and distributed all over Eldan.

Table 5.1 Contextual pages linked to *The Inner Circle* online story

Category of story interactivity	Linked webpages
Intrinsic	Maps
	Glossary
Complementary	Dragonroot Cantina
Adjunct	Loot shop
	Author's Letter
	Reliquary (Movie Poster)
Peripheral	Reliquary
	5 The Relic Cookbook
	6 Gxraden Bolipine's Destiny Cards
	7 Kublata Alphabet

The Dragonroot Cantina is an online 'chatroom', embellished with options such as songs, supposedly representative of those sung at such locations in Atracia, and descriptions of drinks also typical of those available in such taverns. This is not directly related to interactive reading of the story but may provide opportunities to discuss some story elements. The Loot Shop at the time of this visit (August 2004) advertises hard copies of the follow-up stories in the Relic Triangle Trilogy, and is hence considered adjunct to the reading of *The Inner Circle* online story. The Author's Letter also provides adjunct involvement since the current version includes, in response to a reader email, a discussion about why the story is written in the present tense. The Movie Poster is an imaginary poster indicating which current movie stars might be cast in a movie of the story. This kind of adjunct activity could be attractive to students. They might work independently and/or in groups to devise competing posters with different casts and discuss why particular actors would be most appropriate to particular roles. The rest of the Reliquary pages are peripheral to interactive story reading, although the Kublata alphabet could be exploited to construct adjunct activities. The planned alphabet wheel can be used to decipher the strange markings on the spine of *The Inner Circle* novel. This kind of resource might then be used to embellish the *Inner Circle* story with student composed additional text using the Kublata alphabet.

Hypertext narratives

Hypertext narratives are those that do not use images in the construction of the story, although some include minimal accompanying images and icons to identify story elements. These are stories essentially communicated by verbal text. The hypertext links are sometimes from individual words within the text and sometimes from pages or sections, and in the latter case they are often activated by clicking on icons. There appear to be very few such hypertext narratives specifically for children and early adolescents. While not designated for a teenage audience, stories on some sites such as Word Circuits (http://wordcircuits.com/gallery/) are suitable for this age group (although teachers need to review stories before recommending them to students to ensure they are appropriate).

About Time (Swigart, 2002) is described as 'digital interactive hypertext fiction' in the form of a novella in which two tales unfold 40,000 years apart with richly thought-provoking and entertaining results. The tales are 'two braided parallel paths – a double helix' communicated through two main characters introduced in a prologue – Cro de Granville and Mouth. Cro de Granville is a public social commentator with links to some form of research institution who obtains his income through dubious public commentary on social phenomena and unfounded but newsworthy pronouncements of new theories. Mouth appears to be a character from a civilization of more than 40,000 years ago. There are indications in the music and border panels on the pages that this is an Australian Aboriginal Community of that time.

The title screen shows the title and author information superimposed on a beachscape background with a large translucent sphere containing 'bubbles' in the foreground and the caption 'Files from the distant present' at the bottom. This page has musical accompaniment. A 'thumbnail' image of the sphere becomes the icon for navigating to the de Granville files and a similar thumbnail of a fish inside a brown square becomes the icon for navigating to Mouth's journey. If you select the De Granville files, you are taken, with musical accompaniment, to a page showing the icon on the top left and 'chapter' titles in a column immediately below it. At the bottom of this column is a dynamic image of a drop (of water?) continuously dripping onto a target. To the right of the icon is the permanent de Granville files banner: 'The de Granville Files – Present Day – Part 1'. Below this is a subtitle, which changes in accordance with the 'chapter' title selected in the left-hand column. And below this subtitle is the main text for the chapter. These texts may contain hyperlinked words. The hyperlinked words shown in blue, correspond to the 'chapter' titles in the column and clicking on them will take you from your current chapter to the hyperlinked chapter. Hyperlinked words shown in red link to a small additional 'window' containing an image and a voice-over comment related to the hyperlinked word. These are usually comments about the character or behaviours of de Granville by another character or a narrator. Clicking on the 'drip' at the bottom of the column transports you to the pages of Mouth's story.

The layout of Mouth's pages are similar to those for de Granville, except that the 'drip' image at the bottom of the 'chapter' heading column is replaced by the de Granville sphere, and the banner for Mouth's pages is 'Mouth's Journey 40,000 Years Ago Part 1'. There are only blue hypertext links in the texts of the Mouth chapters. The story comes in two parts and once certain hypertext options have been explored, a hyperlink appears at the bottom of the main text in the chapters inviting you to proceed to part two. This is organized in the same way as part one. A thin bar appears at the bottom of the 'drip' or de Granville icons providing a hyperlink choice between part one and part two.

The entertainment and the challenge in *About Time* is exploring both the hypertext links within the de Granville Files and Mouth's Journey through the blue hyperlinked words in the texts and the links across these stories via the hyperlinked 'drip', fish and sphere icons. The story site has a facility which enables you to leave the site and when you return you have the option of either resuming where you left off last time (and the history of the path you chose is displayed on screen) or deleting the previous participation history and starting at the beginning again. What remains unclear is the nature and extent of the exploration of the hypertext options in part one that reveals the hyperlinked option to proceed to part two – and you can't start at part two or get to it until you satisfy whatever these participation requirements are. Nevertheless, the sophistication of the hypertext design of this e-narrative is a fascinating digital rhetorical strategy enabling the creation of an innovative story and engaging the participant in new forms of reading practice and narrative interpretation.

The *Glass Snail* (Pavic, 2003) is described on the Word Circuits gallery page (http://wordcircuits.com/gallery/) as 'a haunting hypertext tale of two people brought together by a shared compulsion. Then the past begins to emerge mysteriously through their meeting, until it threatens to subsume the present.' Their compulsion is to steal some minor item everyday and give away this or a previously stolen or other item on the same day. The two main characters become mutual targets in the pursuit of their shared compulsion and this is what brings them together. The reader can choose which of the two introductory chapters ('Miss Hatshepsut' or 'Mr. David Senenmut, Architect') to read first and with which of the concluding chapters to end the story. These choices will influence the line of the story and its ending. It is possible to read the story several times making different choices of beginning and ending chapters, but these are the only navigational choices one has, apart from going forward or backward from one page to the next, once a path is chosen. There are no hyperlinked words within the text. You can leave the story and resume reading from where you left off upon your return or you can delete the reading history and start again from the beginning. In this story you cannot access the chapters separately and if you want to access a particular chapter, you must progress through all of the pages up to that point. The story itself is quite engaging but the use of hyperlinks is limited and hence readers' options for physical interactivity are also limited.

A collection of hypertext stories for young readers, older readers and teens is provided on *The Stacks* website (http://www.coder.com/creations/tale/stacks.html). They include science fiction, historical, detective/mystery, quest and other story types. These are 'choose your own adventure' stories, but they are 'neverending'. Readers are invited to choose from the paths created by others or to add a new path of their own. Many of the stories are very extensive. At the time of writing *The Haunted Castle* had over 6,000 pages. Each page of the tale comes up on the screen to read. At the bottom of each screen are the choices or paths to follow and the option to add a new path. The site has separate pages providing instructions on how to participate. The pages are numbered and there is a facility to 'jump' to any page you want when you begin the story or return to it. These stories are based on a fairly simple and limited application of hyperlinks.

A distinctive feature of many of these hypertext narratives, especially those discussed above from Word Circuits, is the explicit mechanism for tracking and recording hypertext choices and constructing a hypertext reading history. One way of engaging students in discussion about the digital rhetorics and narrative techniques of these stories is to make use of the reading histories of individuals and/or groups of students as a basis for sharing patterns of reading experience of the texts and discussing the narrative impact of the various combinations of choices.

Hypermedia narratives

Hypermedia narratives make combined use of the opportunities within digital texts to deploy images, image/text relations, hyperlinks and windows to construct innovative narrative forms and to provoke 're-creative' interpretive reading. Teaching/learning activities like those discussed in the sections above can therefore be used separately and/or in combination to explore, and develop students' understanding of, the literary 'constructedness' of hypermedia narratives. But many of these multimodal e-narratives adopt different emphases in their recruitment of the various affordances of digital texts in constructing the distinctiveness of the particular story. In this section five very different hypermedia narratives are briefly described and distinctive semiotic features of the digital narrative techniques used are illustrated, to indicate the importance of working with students to develop their understanding of the relationships among digital narrative forms and the interpretations of story derived from them.

Some purpose-written CD-ROM 'talking books', used predominantly in some school literacy programs, have been criticised (Miller and Olsen, 1998) because they result in what has been called 'truncated learning'. This seems to be because the nature of the hypertext links encourages interactivity with the electronic media but with discreet elements that are not cohesive or fundamentally related to significant aspects of the story. On the other hand, some purpose-written electronic story materials have deployed hypertext and multimodal resources in ways that enhance literary construction of point of view and metafictional elements, to engage readers in active, reflexive story reading in ways that would not be possible in conventional book formats. Two such stories on CD-ROMs were used in a case study of children's exploration of electronic narratives (James, 1999). One of these was *Payuta and the Ice God* (Ubisoft, n.d.). This is the story of an Eskimo boy whose sister is kidnapped by Kiadnic the Ice God to be his cook. With the help of creatures like a narwhal, a polar bear, and an eagle, he reaches the Mountain of Clouds where he finds the fallen Ice Star – famous in Eskimo legends. Payuta picks it up, releasing Nature from the Ice God's grip. Spring returns and Kiadnic's power is destroyed. Payuta and his sister return home truimphant. James (1999) describes how a range of hypertext links enhance the story. For example, in the ice cave clicking on a number of hypertext links (hot spots) made Kiadnic's face metamorphose menacingly out of the rock, accompanied by frightening music. In other sections of the story clicking on the illustrations makes the graphics interactive. In one example the perspective changes from viewing Paytuta on a rock ledge to a bird's eye view of the river below as if the viewer were poised on the ledge, aligning the viewer with the hero's point of view. The second CD-ROM, *Lulu's Enchanted Book* (Victor-Pujebet, n.d.), similarly included hot spots integral to enhanced interpretive possibilities of the story. Frequently the written text entices the reader to activate hidden illustrations. For example, at one point the text states 'Lulu loved masquerading in the most outrageous outfits and posed before the mirror.' The mirror is a hot spot that makes Lulu comment aloud on her appearance as she simultaneously changes into one of three costumes. James (1999) further describes the ways in which hot spots are used to enhance the inclusion of metafictive elements. For example, a subtitle 'The Cutout' is a textual clue to the fact that if the reader clicks on the image, one character draws a frame around another character and cuts her out of the page, rolling her up like a poster. These electronic stories blend some aspects of hypertext and linear models of narrative so that we have these different mindsets Hunt (2000) referred to, to some extent co-existing in the one story.

The relationship between linear and hypermedia models of narrative is what Joellyn Rock set out to address in *The Vasalisa Project* (http://www.rockingchair.org) (Rock, n.d.). At the centre of the project is the story *Bare Bones*, which is a new version of the Russian fairy tale, *Vasalisa and the Baba Yaga*. By reshaping the original story's text, imagery and format, Rock indicates that she is attempting to build a bridge for the fairy tale audience between traditional media and new media. *Vasalisa Electric* is her hypertext essay designed to be 'braided' with *Bare Bones*, which sets out her thesis about graphic design, electronic literature and the fairy tale. The third element of the site is labelled *Hot House* and is the portion of the website designed to house audience contributions, both text and images.

The website layout (Figure 5.6) shows, on the left-hand side, a square 'embroidered' chart with white lines and font on a black background with red hyperlinks. Text is located on the right-hand side of the screen. Navigation of the site is via the embroidered chart. Along the top border of the chart is a panel of icons, each representing an episodic element (node) of the story. Clicking on these brings a static or dynamic story image and a hyperlinked name for the story node to the main area of the chart inside the border. (The nodes are 1. GRIEVE, 2. LEAVE, 3. NAVIGATE, 4. ENCOUNTER, 5. SERVE, 6. SEPARATE, 7. INQUIRE, 8. EMPLOY, 9. MORPH.) Clicking on the hyperlinked node brings the main text of that episode up on the

Figure 5.6 Vasalisa site layout

right-hand side of the screen. The main text includes hyperlinked words, which bring different static images into the main area of the chart. The main text also contains additional static images. The left-hand side vertical border panel of the square chart contains the numbers one to nine, which indicate the main sections of the *Vasalisa Electric* thesis, and these are named to correspond with the node names in the *Bare Bones* story. The right-hand side vertical border panel of the chart also contains the numbers one to nine, and these contain the contributions from readers. The bottom horizontal border panel of the chart is a title panel, which displays the name of the element of the site (Bare Bones, Vasalisa Electric or Hot House), which the reader is currently accessing. Any node of the story can be accessed in any order at any time, and can be cross-referenced to *Vasalisa Electric* and *Hot House*.

Vasalisa is one of the multiple Russian versions of the tale of *Cinderella*, and this site also provides a link to 'Cinder Tales'. In this *Bare Bones* version of Vasalisa, a young girl, following the death of her mother must endure the abuses of a stepfamily. Her little doll serves as both transitional object and 'palm pilot', helping her navigate a series of tests and survive an encounter with the powerful hag Baba Yaga. This is an interactive narrative in a number of interesting dimensions. The digital rhetorical structure of the narrative as a whole is essentially linear. The hyperlinks from the nine chapter titles/ nodes on the top border of the chart progress the story lineally, episode by episode, and the hyperlinked words and phrases within nodes elaborate, illustrate, symbolize and in other ways explore visually, the interpretive possibilities of the hyperlinked text. It is possible to read the nodes in any order and to choose, or not choose, the within-node hyperlinks – and in any order, but such departures from lineality would be a resistant reading of this story, where the hyperlinks confirm the linear structure. However, as Joellyn Rock points out in *Vasalisa Electric*, readers of any text assert their own agency in shaping the interpretive possibilities the text offers. They read at their own speed, jumping, skipping, reading backwards ... and interpreting the tale as they wish. 'But some digital narratives provide new opportunities for participants to work hard / play hard at the recreation of fiction.' Rock makes innovative use of language and images, separately and in combination with hyperlinks, to provoke the reader's exploration of the possible 'recreations' of *Vasalisa* stimulated by her *Bare Bones* text.

The language of the story is designed to unsettle the usual expectations about traditional 'Cinderella' stories and the ideologies they assume. At the beginning of the story simple repetition is used to alert the reader to assumptions of the inherent goodness of 'Cinderella' characters:

> She was the sweetest thing,
> a really
> REALLY
> good girl.
> Her mother dressed her in the perfect
> good-little-girl-little-outfit
> with a black skirt and a white apron,
> a white blouse and a red vest
> all embroidered
> and painstakingly
> designed.

Throughout the story the conventional language of the traditional tale is juxtaposed with contemporary colloquial expressions. For example, the text describes the miniature doll in the image of Vasalisa given to her by her Mother on her deathbed:

> Stitch by stitch her mother had made
> this doll
> this gift
> for her daughter
> by hand.
> The mother stroked her daughter's braids
> and handed her the doll.

The response by Vasalisa disrupts the expectations of the continuation of the conventional storying:

> At first, the doll gave Vasalisa the creeps.
> What was her mother trying to do?
> giving her
> this mini self
> a silly finger puppet?

This disruption is also achieved by the interpolation of contemporary verbal imagery into traditional description, such as, '... her long braids twisting like DNA down her back'. An extension of this 'counter-expectation' strategy is the use of a clause structure that includes three grammatically similar elements that could plausibly be interpreted either as additional or alternative elements within the context of the sentence. For example, following her mother's death, Vasalisa was often to be found

> sitting in the dirt
> behind the apartment building
> under the jungle gym
> beneath the tree near her mother's grave

This partly establishes the conflation of the traditional and contemporary settings of the story but it also raises the possibility of reading the locations as additional or as alternatives. This kind of unsettling, provocative form of 're-creative reading' is taken further with subsequent examples of this structuring as the story progresses. When Vasalisa is on her way to see the old hag, Baba Yaga, we read:

> Suddenly she was blinded
> by the white of some headlights
> and looked up to see
> a station wagon
> a delivery van
> a horseman

Later Baba Yaga lists the tasks that Vasalisa must perform:

> 'You must pick out:
> the poppy seeds from the dirt
> the periods from the pixels
> the dashes ---- • • • • from the dots'
> 'And over here …
> See this pile?'
> Vasalisa looked at an even larger mound
> 'You need to SPECIFY these for me:
> the good grain from the rotting wheat
> the CMYK Gold from the RGB yellow
> the elegant typeface from the grungy font'

The images are also designed to optimize the agentive role of the reader in the re-creative construction of the narrative. With only one exception, the images are highly schematic. They are not realistic or naturalistic in that they are essentially black and white line drawings, which are either silhouettes or outline drawings with only generic representation of human features. From a naturalistic point of view, the images have low modality. This is especially the case for Vasalisa (while we do see a stereotypical witch-like outline drawing of Baba Yaga). This leaves it up to the reader to construct an imagined visual representation of Vasalisa and the other characters. The exception to the schematic, low modality depiction is the doll, given to Vasalisa by her dying mother. The doll is depicted more realistically in full colour. It is also depicted as a very small image, consistent with its story description as being very tiny. The contrast of the doll with the other images affords it salience, drawing attention visually to its role in the narrative.

The low modality of the images is not to say that they are inert with respect to constructing interpretive possibilities, leaving all interpretation to the imagination of the reader. The active role of the images in constructing particular interpretive stances can be seen in the images hyperlinked to the text elements 'took her to bed' and 'listen to her' in the first node (GRIEVE). Both of these images are long distance, high angle, oblique views of Vasalisa kneeling at the bedside of her dying mother. Such images position the reader as not only being remote and detached from the represented participants but also being in a powerful position with respect to them. There is no direct engagement with the gaze of any character, nor any intimate close-up views, so the reader is constructed as having interpretive power and standing outside the interactions of the represented participants. The reader, then, is constructed by the choices of images as a narrative 'operative', re-creating the story from the multimodal representations.

The inter-modal (image/text) relations also stimulate the agentive role of the reader. One example of this is in the NAVIGATE node, which begins:

Never,
ever
had Vasalisa
been
so scared.

The 'N' beginning 'Never' is a large stylized letter with arrows at the end of each stroke separated in the wide lines of the letter by intricate patterns of what later are revealed as light bulbs. Further on this text describes Vasalisa's scary route seeking Baba Yaga:

> As Vasalisa traveled
> she passed many strange
> SIGNS
> (some more readable than others)
> There were
> LOGOS
> and symbols,
> Posters
> banners
> and
> NEON
> marquees
> left by others
> who had marked this trail
> in other eras before hers.

In this segment of the text NEON is hyperlinked and the image that appears in the centre of the chart is a dynamic version of the 'N' that began the chapter, with the internal patterns of light bulbs flashing like a neon sign. This is not only visually clever, but it also activates the intertextual reference to the beginning of the node when 'Never ever had Vasalisa been so scared'.

There are many dimensions of interactivity to be explored in this story, not the least of which are its metafictive elements. One example occurs in the description of Baba Yaga:

> And those legs!
> one all bone with spidery veins
> and the other ...
> lumpy and brown,
> made out of excrement.
> Excrement?
> (that's what it says.)

From the point of view of the affordances of digital rhetoric in constructing electronic narratives, a distinctive message from the experience of reading *Vasalisa* is that a high degree of interactivity can be achieved even when the hyperlinks confirm rather than subvert the linearity of the story structure.

A very different kind of e-narrative on the Eastgate site (http:// www.eastgate. com/LastingImage/Welcome.html) is *Lasting Image* (Guyer and Joyce, 2000), set in Japan in the time just following the Second World War. In this story the interactivity is primarily achieved through a range of different kinds of hyperlinks. The prologue describes *Lasting Image* as offering 'twin views on a single story ... whose link portals open to each other and extend outward to tangential but associated versions'. Forward and back arrows appear on the bottom right corner of each of the main text pages. Further links are located in segments within the images. On the 'waterpaper' views these links are signalled by demarcated, sharply focused image segments against the blurred appearance of the main image. In the clear views the hyperlinks within the images are not overtly signalled and one can only find these by moving the mouse over the image. There may be several different links within the one main image. There are also invisible links in the main texts, again only found by moving the mouse over the text. It is the non-linearity achieved through the range of inter-related different kinds of links that characterizes this story. The experience then, as described in the prologue, is of moving 'between clarity and ambiguity along the blurred margin where control and lack of control flow together'.

Yet another kind of hypermedia narrative on the Wordcircuits site (http:/ /wordcircuits.com/gallery/childhood/index.html) is *Childhood in Richmond* (Zervos, 2001). This story is about a young man's recollections of his father and mother and their fish shop in Richmond. It opens with a large pulsating drawing of a housefly constantly moving from foreground to background against a backdrop of changing/flashing blurred black and white images of boys and a man and woman, apparently represented as a family. There is a light background of drumming music. This simply continues until one moves the mouse over a segment of the screen on the left-hand side. Then a voiceover says 'I remember my childhood in the backstreets of Richmond' and the text 'Childhood in Richmond' is superimposed on the screen. Invisible hyperlinks are located in various areas of the image and passing the mouse over the images activates further voice-overs when the hyperlinks are encountered. A different category of invisible link requires the clicking on some specified area of the screen and this will bring a new graphic of flashing images of people and scenes within the vicinity of the Richmond fish shop and a further range of potentially activated voice-overs. Some of these return to prior screens and voice-overs. This kind of multimedia and hypermedia structuring seems particularly apposite to the narrator's mediation of childhood memories of family and life in the fish shop in Richmond.

Even these few examples are testimony to the rich complexity of digital narrative techniques in emerging hyperfiction for children and adolescents. Notwithstanding the limited work on poetics or stylistics in e-literature, these hypermedia narratives are deploying for literary purposes the digital social semiotic resources that young people encounter in their lives in cyberspace. If literary experience is to be widely nurtured in the cyberspace lives of future generations, knowledge about, and engagement with, the digital techniques of e-literature needs to be taken into account in English and literacy teaching in school contexts.

Digital poetry

In the introduction to this chapter e-poetry was distinguished from hyper-poetry. Central to the meaning, engagement and appeal of e-poetry is its dynamic nature. It is the movement that is constructing the options for meaning making – as well as the nature of the images and the language.

Xylo (Howard, 2001) on the Wordcircuits site (http://wordcircuits.com/gallery/xylo/index.html) is an example of animated e-poetry. In this dynamic poem 'crosshairs' move around a screen and words in different colours appear and disappear. Then segments of the continuous text of the poem appear and disappear in different locations. This continues for about a minute or so until the dynamic poem has run its course. Although a great deal of appreciation of the art of dynamic e-poetry can be developed through viewing and re-viewing the dynamic sequence, it is also useful to consider the use of screen capture software to visually 'grab' selected moments as image files to reflect in some detail and on the effects of the visual and verbal semiotics in the poem at different points in its progress. These kinds of 'disruptive' analytic explorations could be productively undertaken by students and images stored and re-viewed as a basis for enhancing discussion of the poem.

Different kinds of hyper-poetry lend themselves to different kinds of exploratory learning activities. In some cases hyperlinks and 'roll overs' are used to construct patterns of mini dynamic sequences and these may be usefully explored also using the kind of screen capture strategy suggested above.

Stained Word Window (Larsen, 1999) on Wordcircuits (http://wordcircuits.com/gallery/stained/index.html) seems to fit into this category. This poem shows a blue octagon with a purple 'X' superimposed. Coloured words are printed horizontally and obliquely in the segments of this composite figure. By 'skating' the mouse over the words in the figure the segments of the poem corresponding to the hyperlinked words appear on the right-hand side of the screen. Further poem segments can also be accessed by clicking on hyperlinked words within the text segments of the poem as they appear.

Other examples of hyper-poetry include explicit mechanisms for tracking reading history in selecting different hyperlinked options in the poem (similar to those described in the section on hypertext narratives above). An example of such poetry is *The Ballad of Sand and Harry Soot* (Strickland, 1999) on the Wordcircuits site (http://wordcircuits.com/gallery/sandsoot/index.html). This is a mysterious poem presented over 33 pages. The background for each page is black and the poem segment on each page consists of text of varying colours and a small abstract image, of which some are dynamic. There are three navigation methods through The *Ballad of Sand and Harry Soot*, and the reader may combine them in any fashion. They are:

- random reading;
- complete reading;
- link-driven reading.

In the 'random' method the reader chooses any grey zero from the navigation bar at the bottom of the page in any order. Zeros turn white when the pages they link to are visited. The 'complete' reading approach simply means that readers click on each image until they return to the beginning of the poem. (On the one page where an image covers the entire background, readers need to click on the verbal link 'ZaumZoom'. This is made explicit in the advice about possible navigation approaches.) Link-Driven navigation involves the reader choosing how to respond to the two link words on every page. (The only exception is in the case of ZaumZoom, a third link word.) Links will not appear underlined and will sometimes be the same colour as surrounding text. Moving the mouse over the words reveals the links. As noted in relation to hypertext narratives, the facility for noting an individual's and/or groups of students' reading histories in navigating the hyperlinks provides a fascinating basis for student exploration of the interpretive possibilities of the various versions of the poem.

E-comics

Any doubts about the literary significance of contemporary comics, or graphic novels as they have become known recently, especially in their extended form for older readers, would soon be allayed by the innovative scholarly work of Scott McCloud (1994; 2000) examining the narrative techniques of comics and their literary and social significance. The web has been a boon to comic publishing and to innovation in the narrative form of comics. Online comics range from free, static, black and white drawings such as the stories of *Sticky Burr* (http://www.fablevision.com/place/library/STICKY/index.html) to the coloured framed drawings of *Steve Conley's Astounding Space Thrills* (http://www.astoundingspace thrills.com/daily/index.shtml). The free comics on the latter site are also primarily static but occasionally include dynamic elements in the images. There are a number of other subscription websites (http://www.comicsontheweb.com/, http://www.onlinecomics.net) and these often provide free previews. Here we will focus on the Scott McCloud site (http://www.scottmccloud.com/) as it provides discussion and examples of the ways in which the affordances of the web are being used to extend dimensions of the narrative form of e-comics.

Many of the e-comics on McCloud's site are free. They indicate the range of topics covered in contemporary e-comic form. McCloud's first online comic, *Porphyria's Lover*, is an adaptation of a poem by Robert Browning (http://www.scottmccloud.com/comics/porphyria/index.html). This and other comics on this site indicate how the web is facilitating comic artists' going beyond the conventional pattern of panels read left to right and framed in a rectangle. *My Obsession with Chess* (http://www.scottmccloud.com/comics/chess/index.html) tells the story of how Scott McCloud's teenage obsession led indirectly to his career in comics. The narrative form of the story is as engaging as the story itself. The panels are constructed in the classic black and white of chess, alternating white on black and black on white backgrounds. The story scrolls down over 16 feet (approximately five metres) and involves moving from side to side in a chessboard pattern (on a rich dark and light brown 'wood' look chessboard background). Working out how to read the story provides a good deal of interest in itself. There are many other innovative comic designs. One fascinating technique adopted in McCloud's subscription story for mature readers, *The Right Number* (http://www.scottmccloud.com/comics/trn/intro.html), involves the reader clicking on a miniature comic panel inside the main panel and this makes the emerging frame appear from within the previous panel. This subverts the concept on 'pages' online in favour of a kind of visual tunnelling, with each page successively zooming out at the reader. Students can investigate these techniques by comparing online comics using different approaches and also by comparing online and conventional hard copy comics dealing with similar stories. Authoring software accessible to students also makes it possible for them to experiment with retelling of comics, or indeed their own online authoring. For example, a simple retelling of cartoons can be achieved by copying images and loading them into a Powerpoint file. The images of subsequent pages can be copied, reduced to a miniature size and inserted on top of the previous page. A further full size copy can also be inserted over the top of previous image and one of the Powerpoint 'animation' options selected, such as 'dissolve'. Then in Powerpoint 'slide show' mode, clicking on the miniature frame (actually clicking anywhere on the slide) will cause the subsequent comic frame to appear to dissolve out of the miniature frame. The key aspect of this kind of exploration to emphasize with students is 'why' the effect has been incorporated in relation to the other options for constructing the narrative, drawing attention to the potential narrative impact of different forms of story presentation.

Conclusion

Digital narratives and hyperfiction are beginning to form a significant part of the reading experience of young children, adolescents and adults. Young children are being socialized into the world of e-narratives, in many cases as parents and teachers use electronic texts as a resource to apprentice them into literacy. But older children, adolescents and adults are increasingly reading digital fiction for pleasure. This includes a good deal of literature that was previously published in paper format, as discussed in Chapter 4, as well as the e-literature originated in electronic format, which has been the focus of this chapter. But it also includes a range of new forms of digital narrative experience that are integral to many types of electronic games. The importance of these electronic game narratives is taken up in the next chapter.

Chapter 6

Electronic game narratives
Resources for literacy and literary development

Introduction

The acknowledged development of new forms of multimedia electronic literary narratives (Hunt, 2000; Locke and Andrews, 2004; Mackey, 1999) include many multimedia games which could be considered 'hypertextual literature' (Ledgerwood, 1999; Zancanella *et al.*, 2000). While Gee's (2003) discussion of the implications of video games for learning and literacy does not focus specifically on the literary qualities of these games, he does make some useful observations about the relationships among video game stories and the stories in books and movies, and also the different narrative techniques used to influence the 'identification' of readers/players with video game story characters. Gee's work concerns

> ... the sorts of video games in which the player takes on the role of a fantasy character moving through an elaborate world, solving various problems (violently or not), or in which the player builds and maintains some complex entity, like an army, a city or even a whole civilization.
>
> (Gee, 2003: 1)

These *video games* are predominantly character/activity based. They often have plots that may be quite elaborate, usually involving quests of some kind, but the centrally engaging feature is for the player to be continually pitting the capacities of the characters against the obstacle events and phenomena they encounter. In so doing, the characters may be developed in a variety of ways, and a range of different stories may be constructed through playing the game. Gee points out that: 'There are, of course, lots of other types of video games.' Of these other types, this chapter is most concerned with *electronic game narratives*. Here the game activities tend to foreground engagement with a range of aspects of a story that exists in parallel with the game rather than being composed as a result of it. That is, there is a story, which may have been already published separately or is newly composed, incorporating or constituting the game, and the game activities have various subsidiary roles in relation to that story. The second and third sections of this chapter will address some literary aspects of *video games* discussed in Gee's (2003) book. The fourth section will propose a framework for understanding the different types of *electronic game narratives* and their role in the development of literary understanding. The fifth section considers opportunities for the inclusion of the games dimension in classroom activities with literary narratives.

Video games as 'embodied' stories

As Gee (2003: 81–2) notes, in books and movies the order in which the reader/ viewer experiences the narrative events is determined by the author(s). It may be that the book or movie begins with events from the middle or end of the story and the subsequent events may not be related in chronological order, but whatever the ordering is, it is the same for all readers/viewers and is determined for them by the author(s). On the other hand, in video games different players will make choices that result in their experiencing the events of the narrative in different sequences and different players will derive similar information relevant to the storyline from different sources. As well as this, the players will themselves engage in actions as participants in the story and different players will engage in different actions or the same ones in a different order. Hence Gee refers to video game stories as 'embodied' stories since they are embodied in the player's own choices and actions in a way they cannot be in books and movies. Gee goes on to note that there are many reasons why the stories in video games at this point in time cannot be as rich or as deep as those in books and movies. For example, in video games the range of story developments must be based on the choices different players have made earlier in the game. This creates computational complexity that is obviated in books and movies where the author(s) control the earlier choices. It is also the case that video games cannot accommodate the complexity of real conversation within the current computational power available in the game software. However, the point is that video game stories are not necessarily better or worse than stories in books or movies, but they do offer a different kind of narrative experience, especially in how players relate to characters.

Video games and identification with characters

In most of the video games Gee (2003) discusses, players take on the role of a character in the story. Often they can choose which of the characters they wish to become. This story character is referred to by Gee (2003: 58) as the player's virtual identity. Characters have certain attributes such as physical and intellectual skills and personality traits, and the player can often opt to vary the relative strengths of these for the character s/he adopts. During the game, choices made by the player may also enhance or diminish these abilities and this affects the parameters of what the character can do. Hence the player with his or her real world identity is developing a character through actions taken in participating in the game and establishing a virtual identity through his or her role in playing this character. Gee (2003:58) argues that, to the extent that the player projects his or her own hopes, values and fears onto the character, this virtual identity becomes a projected identity of the player. The relationship of 'player as virtual character' is very different from the identification of readers with characters in novels of movies. This is because the relationship is both active and reflexive. Not only does the player actually do things as the virtual character, but also once s/he makes certain choices, these affect the future of the character, which in turn affects what the player can do in the game. Beyond the outcome of the game, there can be a great deal personally at stake through the operation of the projected identity. This projected identity, created at the intersection of the player's real world identity and the virtual identity of the character, helps to speak to and possibly transform the player's real world hopes, values and fears (Gee, 2003:199–200). Of course, this may not happen at all, just as many readers may not be at all influenced by their reading of very significant literary texts, but it does seem that Gee has identified a mechanism of potential learning through narrative which is distinctive of video games.

The narrative techniques that influence the player to align his or her real world identity with their virtual identity in the game are briefly discussed by Gee. One example is from the game *Tomb Raider: The Last Revelation*. Here the character Von Croy has the role of mentor to the young Lara Croft, who represents the character role assumed by the player. Gee points out that the instructions of Von Croy, ostensibly to the character Lara, are also instructions directly to the player as to how to manoeuvre the virtual character. This is very clear in that the language is the language of computer/game console manipulation and hardly the language that would be normally be used to instruct a colleague:

'The first obstacle, a small hop to test your – how do you say – pluck. Press and hold walk, now push forward'.

(quoted in Gee, 2003: 117)

As Gee notes, it is this melding and integrating of the talk to Lara and talk to the player that supports the alignment of the player's real-world identity as player and his or her virtual identity as Lara (Gee, 2003: 117).

Although not directly concerned with issues of the literary nature of video games, Gee's work provides a useful impetus for further work in this area, but as he noted, there are other types of video games and some of these are quite directly related to literary narratives, and we will now turn to these *electronic game narratives*.

Towards a framework for describing electronic game narratives

The approach taken in proposing this framework, as shown in Figure 6.1, is typological. In other words, I have attempted to set out the game activities as discrete categories. As a starting point this approach is quite useful in order to get a clear idea of the different types of electronic game narratives, but it will also become clear that typologies do not always strictly hold, and in some cases particular electronic game narratives will tend to have some features of more than one of the game activity categories proposed. It should also be noted that this is a tentative framework based on a modest sampling of games and does not necessarily provide an exhaustive account. Nevertheless, it is useful for teachers and students to begin to build an understanding of the options available so that they can think about how different choices influence both the nature of the game and the interpretive possibilities of the narrative.

Game-focused stories

The first element in Figure 6.1 distinguishes the video games discussed by James Gee (Gee, 2003) from the electronic game narratives described here, as explicated in the introduction. Electronic game narratives can then be categorized as *story-focused games* or *game-focused stories*. In the *story-focused* category a complete story exists from which the game has been derived or into which it has been interpolated. In the *game-focused* category the game activities constitute all or part of the story. The *game-focused* category

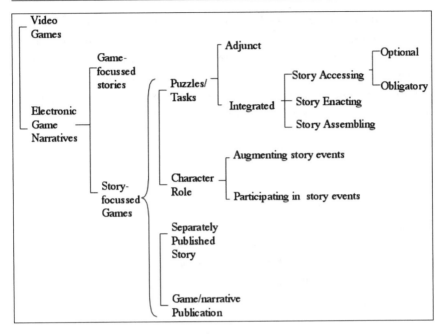

Figure 6.1 A framework categorizing types of game activities in video games and electronic game narratives

seems somewhat similar to Gee's video games, but a number of examples indicate the usefulness of distinguishing this category. One of these is the *Spywatch* game narrative, which was originally published by the BBC to accompany a television series. The BBC have recently taken *Spywatch* offline but educator Ben Clarke has (with permission) made it available for free download for teachers at: http://www.lookandread.fsnet.co.uk/downloads/ sites.html#spy. *Spywatch* is set in wartime England and traces the story of children who are investigating the townspeople to help the police discover a spy who is active in the town. Players practice spy skills, then go and investigate suspects and finally piece together the clues to help Norman, Dennis, Polly and Mary work out what's happening in Westbourne. The site makes good use of images, animations, sounds, hypertext and literacy challenges to unlock doors and solve clues to develop the narrative. It is multimodal, interactive, and combines games with the narrative to produce an exciting reading experience for younger children. Although a good deal of the story construction is based on the game activities, their focus is on the overall story development rather than being based on the attributes of characters marshalled to conquer obstacles to the achievement of a quest. There are other game narratives in this category that are similarly game-focused in that the story does not exist apart from the playing of the game activities (which largely constitute its construction), but where the engagement is clearly with the story outcome. An example for pre-teenagers is the online mystery serial, *Arcane* (http://www2.warnerbros.com/web/arcane/home.jsp), where players work with four characters, each with different skills, reading texts and then searching to 'collect' objects as clues as they direct the characters in their efforts to solve the mystery of 'the stone circle'. For more mature students *Clues* (http://wordcircuits.com/clues/) invites readers to solve the mystery by navigating hypertext links and uncovering clues linked to images, piecing the narrative together. Players are informed that their search will lead down back alleys, empty hallways and wooded trails. They must choose which characters to engage with, for red herrings will keep them from their goal. Their score of clues will be their constant travelling companion in the upper right-hand corner but they are also informed that:

> 'We'll leave aside for now the question of whether winning or losing represents the superior outcome.'

Again, the playing of the game activities constitutes the story, but the engagement is with the construction of the emerging story rather than any prowess in completing the game activities themselves.

Story-focused games

Story-focused games are those that relate to a complete story that has been previously published separately from the game(s) or that has been composed incorporating game activities. Perhaps the best known game-narratives based on previously published stories are the 'Harry Potter' games like *Harry Potter and the Chamber of Secrets* (Electronic Arts, 2002), *Harry Potter and the Prisoner of Azkaban* (Electronic Arts, 2004a) and *Harry Potter: Quidditch World Cup* (Electronic Arts, 2004b). There are many more games based on well-known literary texts such as *The Lord of the Rings: The Return of the King* (Electronic Arts, 2004c), *Alice's Adventures in Wonderland* (Carroll, 2000), *Stig of the Dump* (Topologika, 2000/2001) based on the classic book by Clive King (1963), *Snow White and the Seven Hansels* (Tivola, 2001) and *The Jolly Post Office* (InnerWorkings, 1997) based on the innovative picture book for young children, *The Jolly Postman* (Ahlberg and Ahlberg, 1986). Some story-focused games are included in e-stories that have been newly written to incorporate games such as *The Geode Space* (Street, 1999) and *The Wishing Cupboard* (Hathorn, 1999).

Story-focused games: character role

Story-focused games are of two main types as indicated in Figure 6.1. In one type the player takes on the role of one of the characters and plays as that character. In the other type the player engages with certain puzzles or tasks – but as a player rather than a character. Both of these game types can occur within the same game-narrative. In this section we will consider the character role type. The character role games can also be divided into two sub-types as shown in Figure 6.1. The most common of these is *augmenting story events*. This is where the game-narrative adaptation of the original story includes game activities where the character has to solve problems or overcome obstacles by completing activities that are consistent with, but not actually part of the original story. This is a common feature of the Harry Potter games. In *Harry Potter and the Chamber of Secrets* (Electronic Arts, 2002) for example, Harry is obliged to utilize his resources, manipulate objects and navigate pathways to effect the rescue of Ron Weasley in the flying Ford Anglia from the Whomping Willow tree. In the older format *Stig of the Dump* (Topologika, 2000/2001) game, players take on the role of Barney and have to carry out a number of tasks to assist Stig to prepare his cave for a cosy winter. In *The Lord of the Rings: The Return of the King* (Electronic Arts, 2004c) there are ten 'playable' characters, including Gandalf, Aragorn, Legolas, Gimili, Sam and Frodo. Players can explore Middle Earth and fight vicious battles including Shelob, the Witch King, controlling the battles and the fates of their characters.

The second type of character role activity is where the player is invited, as a player, to become a direct participant in the story world. These game activities occur in the *Mulan Animated Storybook* (Disney, 1998) as indicated in Chapter 4. For example, when Mulan is in battle with the enemy leader, the player is required to assist by selecting melons and 'throwing' them at the guards. Then when Mulan is rescuing the Emperor, Mushu instructs the player to open the door (so that he can obtain more weapons).

Story-focused games: adjunct puzzles/tasks

Game activities that are puzzles or tasks may be adjunct to the story or integrated with the story.

An example of the adjunct type is *The Jolly Post Office* (InnerWorkings, 1997) designed for younger children. This is based on the book *The Jolly Postman* (Ahlberg and Ahlberg, 1986). The Jolly Postman delivers letters in a fairy-tale kingdom to such familiar addresses as 'Mr. and Mrs. Bear, Three Bears Cottage, The Woods'. Every other page is an actual envelope, with a letter tucked inside. The letter to the three bears, for instance, is from Goldilocks, who apologizes for the trouble she's caused and invites Baby Bear to her birthday party. The wicked witch from *Hansel and Gretel* receives a circular from Hobgoblin Supplies Ltd. that advertises such appealing products as 'Little Boy Pie Mix'. In the game the activities are not related to the narrative, although they do involve characters from the book. The child has been left in charge of the Jolly Post Office and can choose whether to send postcards, sort letters, swap stamps on the internet, help customers, mend parcels or personalize stationery. The customers are characters from *The Jolly Postman* book, such as Baby Bear and The Big Bad Wolf (dressed as grandmother) and the Witch. The player helps each customer by weighing the parcels and calculating the correct postage. The player can also send postcards to Jack on his travels around Europe, learning all about maps, flags and different countries, as well as a number of other activities including designing their own stamp, starting a collection in their own stamp album, entering their designs in the World Stamp Book and collecting other children's designs from around the world.

A different kind of adjunct game is the Fox Taming Game, which is on the same CD-ROM as the story of *The Little Prince* (de Saint-Exupery, 2000b), but one has to exit the story to play the game. The game is based on chapter XXI of the book. In this chapter the Little Prince meets a fox who asks the Prince to tame him and the Little Prince asks what 'tame' means. Much of the chapter then deals with issues around the importance of committing time to the development of relationships and the mutual responsibilities of those involved in relationships. The Fox Taming Game is remarkably simple. The player explores the virtual landscape where clicking on objects results in some effect, such as clicking on the fountain which causes the water to flow from it. However, whenever the Lamplighter calls out 'Good Morning', the player must click on the fox's den to meet with the fox in order to learn how to tame him. If successful, the reward is access to a personalized fox diary and stationery. But there is no indication of what is required to tame the fox, apart from clicking on the den to achieve his presence at every morning call by the Lamplighter. The game then becomes somewhat tedious since the exploration of the landscape is quickly exhausted after two or three morning calls – and the fox, even after several additional morning calls shows no sign of being tamed. What is required of the player, then, is simply patience and willingness to persist in summoning the fox every 'morning'. Eventually the fox is tamed, and the considerable patience and persistence of the player in the face of an exceptionally uninteresting game routine is rewarded. Hence the key aspect of this game is the way in which it relates to the central theme of chapter XXI and provokes deeper engagement with the philosophical issues at stake.

Story-focused games: integrated puzzles/tasks – story accessing

Puzzles or tasks can be integrated into the story and these integrated activities can have three main functions: *accessing* the story; *enacting or progressing* the story; and *assembling* the story. We will look firstly at two online stories where puzzles/tasks function to control access to the story or story segments. The first of these is *The Geode Space* (Street, 1999). This is a story about 11-year-old Jenny and her 9-year-old sister Kaylee and their similarly aged friends Shanna and Kerry whose adventure begins with their discovery of some purple crystals in the caves near where they live. On the first screen the readers are provided with the following instructions:

> There are secret animals hidden within Chapter 1. Run your mouse across the page throughout the story to uncover the hidden animals! Some are tiny, so you'll have to look everywhere. There'll be one animal in each of the five pages in Chapter 1.
>
> Write down the first letter of each animal's name because you will need ALL 5 of the letters to figure out the secret CODE WORD to get to Chapter 2!

The Wishing Cupboard (Hathorn, 1999) is the story of a Vietnamese boy, Tan, and his grandmother, who opens The Wishing Cupboard for him one rainy day. Tan is longing for his Mum to come home from her trip to Vietnam. Tan's grandmother knows The Wishing Cupboard will distract Tan from thinking all the time about how long his mother has been away and when or if she will be coming home. The player needs to open the doors in the cupboard by clicking on them to reveal an object of Vietnamese culture. On each occasion this is done correctly, the text appears with the grandmother communicating some aspect of the culture related to the object, and this text is linked to a Vietnamese folk story and further information about the country and the culture. Each drawer must be opened in turn to progress the story and one drawer at the top cannot be opened until all of the others have been successfully opened. This is the drawer that allows the player to complete the puzzle and find out if Tan's mother does come home.

The electronic game narrative of *Alice's Adventures in Wonderland* (Carroll, 2000) includes story-accessing puzzles or tasks at the beginning of each chapter. For some chapters (e.g. chapters 1 and 4) these are electronic jigsaw puzzles. Once the pieces are correctly arranged by using the mouse movements, the animated story will begin. To access chapter 2 one needs to click on a half screen full of hidden faces of young girls to reveal matching faces from the display and to access chapter 6 one needs to guide Alice through the sky avoiding a collision with the birds flying across her path. In all of these examples it is possible to avoid participating in the story-accessing puzzle/task by simply clicking on the 'running Alice' at the bottom of the screen. This 'skips' the activity and provides direct access to the animated story. But this 'skip' option is not always available. In chapter 3, for example, the Dodo bird sets up the 'Caucus Race' and the task of the player is to ensure the Dodo bird gets every creature ready on its mark. This involves clicking on the creatures around the perimeter of a circle at the precise time that the line rotating from the centre of the circle passes over them. There are a large number of creatures to click on and the line rotates quite quickly making this a challenging task to complete. However, it must be completed to access the rest of chapter 3, because there is no 'running Alice' available to skip this activity. Some other chapters also have obligatory story-accessing tasks without the 'running Alice' facility. In chapter 5 for example, one is obliged to help Alice reveal the blue caterpillar on the mushroom before gaining access to the animated story.

Story-focused games: integrated puzzles/tasks: story enacting

The *Alice's Adventures in Wonderland* game (Carroll, 2000) also includes story-enacting or story-progressing tasks. For example, in the first chapter when Alice is tying various aids to gain access to the small door, she encounters the cake as the antidote to the potion she has drunk from the bottle, but the player must use the mouse to reveal the location of the currants on the top of the cake, working out that they spell 'EAT ME'. This task must be completed for the story to progress. A further example occurs in the sixth chapter when the player must click on the correct piece of mushroom that will return Alice to the appropriate size for the story to progress.

The Mulan Animated Storybook (Disney, 1998) contains many tasks that enact or progress the story. For example, in the village we find that the scroll containing the story is incomplete and some strategic clicking is required to find the missing parts. One such missing part refers to the grandmother collecting some lucky charms to bring Mulan good luck in her meeting with the matchmaker. The puzzle is to locate and collect the charms from the screen and give them to the grandmother. The successful completion of this task is necessary to progress the story events, since once you have assisted Mulan in dressing to meet the matchmaker, it is necessary to take the bag of charms from the grandmother and give it to Mulan before she can proceed to her meeting. Similarly, in Mulan's house it is necessary to complete the puzzle to form the image of the dragon on the cupboard to access Mulan's father's armour for her to wear to the army camp. It is also necessary to pack appropriate items into Mulan's saddle bag, avoiding the attempts of the dog to foil the packing by inserting his toy ball. Once the dragon puzzle and the bag packing game are complete the Mulan character announces she is ready to leave.

Story-focused games: integrated puzzles/tasks – story assembling

Snow White and the Seven Hansels (Tivola, 2001) is an animated retelling of Little Red Riding Hood, Snow White and Hansel and Gretel. It is possible to follow the guide arrows and simply experience the animated stories. However, by choosing to pathways other than those represented by the guide arrows it is possible to mix the stories so that Little Red Riding Hood could easily end up with the seven dwarves and Hansel and Gretel might well meet the big bad wolf instead of the wicked witch.

Including electronic game narratives in developing literacy and literary understanding

The inclusion of electronic games in classroom learning experiences relating to literary narratives acknowledges their role in expanding the dimensions of story and their potential to enhance students' engagement with the evolving hybrid nature of narrative in the age of electronic multimedia. This section will outline three areas of potential classroom work relating games to literary narratives. The first is the investigation of games as a new literary genre in their own right – examining how their content and form contribute to the creative possibilities for the player and the enjoyment and satisfaction of participating. The second area deals with the links between the game activities and the story, discussing the types of game activities in relation to the selection of aspects of the narrative such as elements of the plot, characters and settings. The third area for classroom work involves exploring intertextual relations between selected literary narratives and electronic games composed without ostensible reference to these literary works.

Investigating electronic game narratives as literary genres

Studies of video games as a basis for developing and expanding our understanding of literacy learning in the multimedia age have indicated aspects of electronic games that contribute to the creative enjoyment of playing in the context of a literary experience (Gee, 2003; Sefton-Green, 2001; Zancanella *et al.*, 2000). One such aspect is the richness of the narrative scenarios in which the game activities are embedded. For example, the *Lord of the Rings Fantasy World* website (http://www.lord-of-the-rings.org/) includes a free online game, which is relatively straightforward, but includes very detailed preliminary pages with an extended account of 'The Game Story' and the peoples inhabiting Middle Earth. *Arcane*, the online mystery serial of The Stone Circle (http://www2.warnerbros.com/web/arcane/home.jsp) provides the context of the game in two stages. The first is the abstractly animated image/text presentation of the *Prologue* and the second is the extended *Mythology*, which details the story from which the game-narrative emerges. Students can be encouraged to select examples of these narrative scenarios that they regard as successful and identify the features that account for their success. These might include aspects of the content, the images and/or image text presentation and/or the design or dynamic presentation effects.

Reflection upon such aspects of the games will be a readily accessible activity for most students, as will some features of characterization such as the extent to which characters' skills and traits can be varied and/or developed as part of the game. Some features of narrative approach, such as whether the game is constructed in a first-person or third-person orientation, can also be relatively easily dealt with in classroom work. However, understanding other narrative aspects such as Gee's discussion of the melding and integrating of 'mentor' character's communication to the game character role assumed by the player and simultaneously to the real world player in relation to manoeuvring game controls (see 'Video games and identification with characters' above), may well require initial teacher modelling and explicit teaching for some students. The introduction of students to this kind of metasemiotic understanding can facilitate their capacity to conceptualize and communicate the basis of their appreciation of the effects of particular aspects of game design.

The framework for describing different types of electronic game narrative activities provided in Figure 6.1 can also become a useful basis for beginning to develop students' explicit metasemiotic understanding. By categorizing different game activities and thinking about how the different types contribute to the overall game narrative experience, students can begin to participate productively in the critical appreciation of these genres. Students may, for example, consider the contribution of the adjunct game activities 'Wizard Duelling' and 'Quidditch' included with the *Harry Potter and the Chamber of Secrets* game narrative.

Relating electronic game activities to aspects of the literary text

Work with students in relating the electronic game activities to aspects of the literary narrative can assist in developing their critical appreciation of the potential role of different kinds of game activities in expanding the narrative space and interpretive possibilities of the story in a range of different ways. This has been noted above in the example of *The Fox Taming Game* and *The Little Prince*. This example clearly indicates that adjunct game activities are not necessarily peripheral to the fundamental thematic concerns of the story, although this appears to be the case in games like *The Jolly Post Office* and the *Quidditch* and *Wizard Duelling* activities in *Harry Potter and the Chamber of Secrets*. An interesting task for students then, taking the *Fox Taming Game* as an exemplar, would be to think about adjunct games that could emphasize key themes of the stories. It may be useful for some students to engage in this kind of work collaboratively with the teacher in the first instance. For example, the class might consider the kind of adjunct game that could be generated around the role of characters met by *The Little Prince* before he visits the

Earth. In the case of characters like the king, the businessman and the geographer, despite their power, wealth and authority, it is what is missing from their existence that is central to the developing themes of the book. One approach to an adjunct game therefore might be the jig-saw construction of each character such that in the case of each character one of the pieces correctly fits the shape of the puzzle but the content and colour of the piece does not match the character. The shapes of these mismatched pieces might be such that they construct a silhouette of the Little Prince and his flower. The goal of the game is to construct the jig-saw of the Prince and his flower, but this cannot be done until the component pieces are identified through the completion of the other characters. Like the *Fox Taming Game*, the process of the game activity is intended to provoke the players to think about aspects of the theme(s) of the story. Following this kind of teacher scaffolding, students might be asked to plan adjunct games appropriate to emphasizing particular themes in the Harry Potter stories and/or other game narratives.

In a similar manner the role of story accessing games might be investigated. In *The Geode Space* story described earlier, the picture/word puzzles determining access to subsequent chapters might be considered somewhat gratuitous impositions in terms of their relationship to the content and form of the narrative. However, in *The Wishing Cupboard*, delaying Tan's progress in exploring the cupboard, and hence the story-accessing game activity, was clearly functional in terms of his Grandmother's concern to distract him from thinking about the length of time elapsing while his mother is away. Students might be asked to consider if, and in what ways, the story-accessing game activities in *Alice in Wonderland* might be considered functional in relation to emphasizing key issues in the story. For example, it might be suggested that since Alice is constantly needing to solve problems about the ways she is to adopt the appropriate size to progress in her journey through Wonderland, the story-accessing puzzles emphasize this theme of the accommodations required in interacting with others. Another possibility is to see these story-accessing puzzles functioning as a means of narrative foreshadowing with the jig-saw puzzles, for example, indicating the key participants and contexts of the forthcoming events. Again, following this kind of scaffolded work students might undertake the planning of additional story-accessing game activities that would enhance readers' engagement with key themes in other game narratives.

The same kinds of explorations could be undertaken with story-enacting/progressing game activities. It may be that in the context of the action-oriented progression of the game-narratives, one role of story-enacting/progressing activities is to ensure attention to details of description and circumstance that help to maintain the richness of context of the game. It would be interesting to examine the story/enacting game activities in *Mulan* from this perspective.

In considering the character-role aspect of game-narratives, students could think about re-designing *Harry Potter and the Chamber of Secrets* as a third person game with progress determined by negotiating integrated story-accessing and story-enacting puzzles/tasks. Alternatively, they could consider what games like *Alice...* would be like redesigned as first person games, so that the player progressed as Alice through Wonderland. These kinds of learning activities are not out of tune with common practices of asking students to rewrite story episodes from the perspective of a minor character or adding an alternative episode to a story. They are very much in tune with the creative 're-storying' contexts that students increasingly see as a part of the narrative experience of contemporary stories.

Exploring connections between literature and ostensibly unrelated electronic game narratives

We know that video games are a significant part of the lives of an increasing proportion of young people from the early years of schooling to adulthood, and notwithstanding that, literature for children and young people in book formats maintains its popularity as reflected in the phenomenal commercial successes of J. K. Rowling's Harry Potter books. A number of case studies (e.g. Davidson, 2000) suggest that many avid game players are also avid readers, and recent research (Mackey, 2002) indicates that children's preferences among stories do not privilege book, electronic or game formats. Clearly then, classroom work with literary texts needs to incorporate cross-media discussions of novels and picture books and new genres of electronic game narratives.

As the acceptance of the role of electronic games in classroom contexts grows, so will the repertoire of choices for literature-oriented, inter-media, intertextual learning activities. One way in which teachers can facilitate this kind of development is initially to make use of well-known novels and picture books and electronic games that have obviously similar thematic concerns. In working with young children in the early years of schooling this might take the form of including work with computer games such as *Arthur's*

Absolutely Fun Day (Mattel Media, 2000) based on Marc Brown's Arthur stories (e.g. Brown, 1999) or online games such as those based on William Joyce's (1985) *George Shrinks* at the pbs kids site (http://pbskids.org/georgeshrinks/gallery/index.html).

Dreamwalker by Isobelle Carmody and Steve Woolman (2001) is an illustrated book that will appeal to pre-teenagers and young adolescents. The narrator, Ken, is a lad who has been an insomniac since childhood. His driving interest is drawing and his ambition is to become an artistic composer of comic books. His long waking hours in the night are largely spent drawing stories and on one such occasion he creates the sorceress, whose vampire-like minions, which emanate from her dreams, feed on unwitting sleepers. One day Ken wakes from a dream and finds himself in the world he has created, where one of his characters, the beautiful Alyssa, claims to have imagined him in a dream. As the intricate plot unfolds, we are drawn into an incredible situation where it becomes impossible, even for Ken, to tell who is the creator, who has been dreamed and who is the dreamer.

A search of the web located *Dreamwalker: Roleplaying in the Land of Dreams* (http://dreamwalkerrpg.home.att.net/). The site indicates that in this game you play a Dreamwalker in the employ of Project Dreamwalker. Each night you enter the dreams of others and attempt to rid troubled minds of the Taeniid infestation. The intertextual possibilities suggest a fruitful opportunity for cross-media study, although it is suggested that, as a precaution, the use of this resource be managed by the teacher as there is no indication that this site or the game is rated as suitable for independent participation by children.

An alternative for readers of this age and for older readers is the new (soon to be released) Sherlock Holmes game, *Secret of the Silver Earring* (Ubisoft (in press)), which could be studied in combination with one or more of the Sherlock Holmes classic novels by Sir Arthur Conan Doyle such as *The Hound of the Baskervilles* (Doyle, 1982). A further possibility is reading the novel *Z for Zacariah* (O'Brien, 1998) in conjuction with the popular video game *Final Fantasy X* (SQUARE-ENIX, 2002). In *Z for Zacariah* (O'Brien, 1998) Ann Burden, the lone survivor of a nuclear holocaust, is threatened by the arrival in her valley of an unknown intruder. She hides, he watches and they both wait. Ann wonders whether this scientist in a radiation-proof suit is an ally or whether the horrors he has witnessed turned him into something more sinister. The answer unfolds in a story that involves a battle of wills, which ends in a struggle for survival between a girl and the last man on Earth. *Final Fantasy X* involves Tidus, a champion of the sport Blitzball, whose world is destroyed while he is in the middle of a match. This story begins with Tidus, the sole survivor, being transported to another world. Here he meets Yuna who has learned the art of summoning and controlling aeons, powerful spirits of yore. These two people of different backgrounds must work together as they journey through the new world of Spira.

The comparative study of picture books and novels and electronic game narratives has the potential to become a very engaging and intellectually rewarding learning experience with literary texts. It is a context for developing literary understanding that acknowledges the contributory role of the multimedia expertise of our children and optimizes their interaction with the more extensive literary experience of their teachers.

Conclusion

Electronic game narratives not only exemplify aspects of the new literacies required to function in the textual habitat of our contemporary digital communication environment, they also exemplify a new form of narrative art, worthy of serious attention in literary studies and in the formal educational experience of children and young people. Game narratives are a heterogeneous phenomenon entailing an extensive and complex range of genres including many different types of game activities. Many of these games have been designed as an integral part of the expanded realm of literary works published as books and movies – often in multiple formats. These hybrid inter-media forms of narrative experience apply not only to emerging new literature but also to the re-presentation of classic and traditional literature. While many games are commercial products accessed via compact disks or by paying for internet downloads, a not insignificant minority can also be accessed gratis via the web. Not only are the games popular with children but they also attract a very large clientele among young adults. Because of their growing social significance, electronic game narratives are not simply useful but in fact imperative as a resource for literacy and literary development in today's classrooms.

Chapter 7

Practical programs using e-literature in classroom units of work

Introduction

This chapter focuses on the practicalities of planning classroom work for children. The facilitative knowledge about e-literature needed by teachers to undertake this challenging pedagogic work has been systematically addressed through the frameworks introduced in Chapter 1 and developed and illustrated in the following chapters. Throughout these chapters the focus has been on the professional knowledge required for resourcing effective classroom practice. Suggestions for teaching and outlines of learning activities related to particular narratives have been interpolated into each of the chapters. In this chapter the emphasis is on planning and implementing cohesive, engaging programs of work encompassing close, reflective, analytic study of some stories and wide, exploratory reading of others, while meeting the expectations of an English curriculum in teaching a class of about 25 students.

The range of narratives explored in this book has included stories suitable for very young and inexperienced readers, such as *George Shrinks* (Joyce, c.1994), to those oriented to more experienced young adolescent readers such as *The Vasalisa Project* (Rock, n.d.) and the Philip Pullman trilogy of *His Dark Materials* (Pullman, 1995, 1997, 2000). Since a good deal of detail is required to establish the practical viability of an extended unit of classroom work, it will be necessary to restrict the range of the program description here to two broad age or class groups. The first program will be suitable for most children in about the third year of schooling or about 8 or 9 years of age. This program is based on Oscar Wilde's *The Selfish Giant* (Wilde *et al.*, 1986 (1888)) in an e-literature context. The second program has been designed for children in about the fifth or sixth year of schooling, usually from about 10 or 11 years of age. However, this program is based on *The*

Little Prince (de Saint-Exupery, 2000a) with augmenting narratives like *Skellig* (Almond, 1998), and hence could well extend to work with students in their early teens. What is presented here is not intended as a prescriptive 'template' for classroom implementation. What is intended is a sufficiently detailed plan of ideas for the classroom that indicates to teachers planning programs for their own classes how e-literature can be practically incorporated.

The Selfish Giant

Oscar Wilde's story of *The Selfish Giant* continues to generate a variety of new illustrated book versions (Wilde, 2001), seemingly countless versions on the web, and a video that can still be purchased 'new' from Amazon.com (Zander, 1992). This story remains popular with many teachers and children and has been a focus for work published on the web by children in the early years of school (http://www.burgh-by-sands.cumbria.sch.uk/selfishgiant.htm). This section will outline possible learning experiences for a program of classroom work based on *The Selfish Giant* for children in about the third year of school. Firstly, we will note some well-known book versions of the story and some of the web resources we could make use of, as well as the broad approach and goals of the program. Then we will outline the actual learning experiences and suggestions for their implementation in the classroom.

The 'Picture Puffin' version of *The Selfish Giant* (Wilde *et al.*, 1986 (1888)) is widely available in school and public libraries in Australia and in the United Kingdom. The images by Michael Foreman and Friere Wright show some menacing close-up views of the giant at the beginning of the book and gentle images of the older giant at the end of the story. In North America the Putnam Penguin version illustrated by Saelig Gallagher (Wilde and Gallagher, 1995) is advertised by Amazon.com with an accompanying audio download narrated by David Ian Davies. Among the many other versions is a 'mini classics' publication (approx. 10 cm × 8 cm) illustrated by Susan Neale (Wilde and Neale, 1994). In this little book the story seems to have more of a late nineteenth/early twentieth century setting. Although there area many versions of *The Selfish Giant* on the web (e.g. http://www.planetmonk.com/wilde/index.html), very few of these include illustrations. The Storysocks site has an opening image of what appears to be a contemporary garden (http://pages.zdnet.com/storysocks/library/id52.html), and the Imaginarium site (http://www.cornerstonemag.com/imaginarium/features/selfish.html) shows the cover of a version of the book illustrated by P. Craig Russell. The 'Fireblade Coffeehouse' site (http://www.hoboes.com/html/FireBlade/Wilde/Giant.html) provides a straight text version of the story, but it also includes an opportunity for readers to post their responses on a comments page. A Readers' Theatre script for the story has been provided by Thomas Eastland on his 'Theatrical Resource Home Page' (http://users.chariot.net.au/~ghost/selgi.htm). The Penguin Readers page (http://www.penguinreaders.com/pyr/resources/index.html#top) provides a free, 'download' pdf file of 'Teachers' Notes'. These include a story summary, suggested themes to explore (buildings, friendship, size and weather/seasons), suggested small group writing of pseudo newspaper reports of events in the story, three chants to accompany the story, a true/false comprehension test, a drawing of the garden for children to label, a crossword and a find a word puzzle, as well as 'adjective' games to familiarize children with vocabulary relevant to size. Although the web provides a very useful range of resources to support classroom work with *The Selfish Giant*, the book versions are essential to enjoy the visual interpretations of the story.

The intention of the classroom program outlined here is to complement the literacy development orientation (reflected in the Penguin Teachers' Notes indicated above) with experiences designed to use the resources of the web to enhance children's literary understanding and enjoyment. The main focus will be on the narrative role of images. The practical means by which children will be engaged in learning about the meaning-making resources of images, and the effects of different types of images in stories, will be based on the approach taken by Monica Edinger with her year four students at Dalton Elementary School in New York (http://intranet.dalton.org/ms/alice/alice.html), which was described in Chapter 4. A key goal, then, will be to support the children in developing their own class website showing their illustrated version of *The Selfish Giant*.

To begin the program, learning experiences will enable children to share their existing experience of stories about giants. The story of *Jack and the Beanstalk* is likely to feature prominently here.

New stories involving 'giant' characters, such as Roald Dahl's *The BFG* illustrated by Quentin Blake (Dahl and Blake, 1982), will also be introduced. *The BFG* (Big Friendly Giant) introduces a different perspective on giants, as he is bullied by his giant peers because, rather than eating small children, his diet consists of disgusting 'snozzcumbers', and he spends his time blowing happy dreams through the windows of little boys and girls. Another distinctive picture book dealing with the concept of giants is Libby Gleeson's, *Uncle David*, illustrated by Armin Greder (Gleeson and Greder, 1992). In this story, the smallest child in the kindergarten class claims that his uncle is a giant, and his little friends quickly spread the news. The more the stories are repeated, the more they grow, but once everyone meets Uncle David, in his small university college study-bedroom, they happily discover that he is not too big or famous or frightening to visit their school for a show and tell session about giants. Other stories, picking up on different themes in *The Selfish Giant*, will also be included in the program. One example that is quite apposite is *Rose Meets Mr Wintergarten* (Graham, 1992). In this story young Rose Summers and her family move to a new house and Rose is immediately curious about the spooky house next door where 'mean and horrible' Mr. Wintergarten lives. The word among the local children is that he has a 'dog like a wolf' and a saltwater crocodile 'that he rides at night that GETS YA!' When Rose's ball goes over Mr. Wintergarten's fence, she ventures to retrieve it, but with her mother's help she also takes cakes and flowers to the old man. Mr. Wintergarten's gruff attitude masks the fact that Rose's kind gestures have begun to melt his heart, and soon the neighbourhood has a new friend.

This program of classroom work assumes the common practice in Australian primary schools of allocating the first session of the school day to work in English/literacy. The duration of these sessions is approximately 90 minutes to 2 hours. A variety of whole class, group work and individual activities are undertaken during this time.

Week 1: Session 1

- The teacher conducts a shared reading of Libby Gleeson's *Uncle David* (Gleeson and Greder, 1992).
- Following this, the children are invited to talk about other stories they know about giants.
- The class is then divided into six groups of about three or four children. Three of the groups of more able readers undertake a 'jig-saw' group work task: each of the three groups is allocated a different *Jack and the Beanstalk* book to read, and half way through the time for this section of the lesson they are re-grouped so that each new group contains at least one member of the original group. The new groups therefore contain at least one 'expert' on the story version read by the original group. The

task of the new groups is to share their understanding of the original story and how the language and the images are the same or different. (Children can record words or phrases that occur in more than one story. They may also be supplied with a bank of adjectives describing the giants and use 'post it' notes to attach these to various images. Then they can list the 'giant' images, e.g. Book 1 Image 1, 2 etc. and Book 2 Image 1, 2 etc. and note where their allocation of a 'post it' adjectives are similar or different for the various images.) The three groups of less able readers are each allocated different tasks. One group works with the teacher to ensure they are able to read and understand the story versions that are being explored in the 'jig-saw' task by the first three groups of children. The second group reads selections of online versions of *Jack and the Beanstalk* such as the one provided by Antelope Publishing (http://www. antelope-ebooks.com/childrens/CHILD/TALES/TALE01. HTM). This version can also be purchased as a browser-readable CD-ROM. The third group uses a graphics program or a free online sketch pad (http:// www.storybookonline.net/article.aspx?Article=Sketch_Pad) to draw two different images of a giant they remember from one of the giant stories they know. These three groups rotate around these tasks so that the teacher is able to give intensive support to all groups in turn.

- To conclude the session the teacher begins a serial reading of *The BFG* (Dahl and Blake, 1982).

Week 1: Session 2

- The teacher reviews with the children the story of *Uncle David* from yesterday and re-reads it with children invited to join in on allocated reading parts.
- The group work tasks from the previous session are 'swapped' across the two halves of the class, so that the jig-saw group now reads the web stories, makes the two sketches of their giant character, but their reading of book versions is an extension to further *Jack and the Beanstalk* tales. The teacher focuses his/her attention primarily on supporting the second three groups to complete the jig-saw task.
- Display the children's drawings of their giants and involve the children in articulating descriptions of them.
- Discuss as a whole class the variety of 'giant' images in the different books they have now all read. Focus on articulating how the images are

the same or different and initiate discussion of how these features influenced what the story was like.

- Continue serial reading of *The BFG*.

Week 1: Session 3

- The teacher signals that today the class will be introduced to a new and different story about a giant and provides a model reading of *The Selfish Giant* from a version without illustrations. The teacher should record his/her reading of the story.
- Discuss the story with the whole class rehearsing the event sequences in the story, commenting on the characters and discussing the lesson(s) to be learned from the story.
- Organize the class into four groups of about five or six children. Try to pair more experienced and competent readers with friends who experience difficulty with reading. Each group is allocated one of the following tasks and then, after about 20 minutes, the groups rotate, so this segment requires about 75–80 minutes. The first task involves the children preparing a group oral reading of the story with children allocating themselves different segments of the story to read. Some groups will do this quite independently and for other groups the teacher will need to assist some children in rehearsing their reading segments. Some peer support may also be possible for this task. The teacher's taped reading of the story can be available on a 'listening post' as a resource to support this group work. The second task involves the children completing some of the activities from the Penguin Readers page (http://www.penguinreaders.com/pyr/resources/index. html#top). Specifically these are activity 1, true/false comprehension questions, activity 2, labelling the illustration of the giant's garden, activity 3, cloze crossword, and activity 4, find a word. The third task involves visiting the 'Fireblade Coffeehouse' site (http://www.hoboes.com/html/FireBlade/Wilde/Giant.html) and posting some initial comments about the story. Again, some students will need teacher and/or peer support in doing this. Comments will be copied to a Word file for review later in the classroom program. The final task is paired reading of *Rose Meets Mr Wintergarten* (Graham, 1992).
- Groups report to the whole class on the comments they wrote online and on their reading of *Rose Meets Mr Wintergarten*. Children are

encouraged to practise their reading parts for a group presentation of their reading of *The Selfish Giant* in tomorrow's session.

- To conclude, a brief recapitulation of *The BFG* story to date and a shortened serial instalment for the day.

Week 1: Session 4

- Serial reading of *The BFG*.
- Rehearsal time for groups of children to practise their reading presentation of *The Selfish Giant*.
- Two groups present their readings to the class in turn.
- Whole class discussion about the images of giants that might be used in this story. Would any of the images from any parts of the *Jack and the Beanstalk* story be suitable for any sections of *The Selfish Giant*?
- The teacher displays at least six different illustrated versions of *The Selfish Giant*. Six groups of children are each allocated a book to read and then the jig-saw group procedure is used to generate new groups where each member has read a different version. (The audiofile download and/or the teacher's recorded reading can be available with headsets to assist students who have difficulty in negotiating the text.) The children informally compare and group the different kinds of depictions of giants. The teacher makes photocopies of the range of giant images and student groups sort these and label with their own categories.
- Whole class discussion and report on giant image categories. A consensus grouping is pinned to a chart in the room.
- Second presentation of prepared group reading of the story by the remaining two student groups.

Week 1: Session 5

- Serial reading of *The BFG*.
- Review the illustrated versions of *The Selfish Giant* read in the previous lesson and the children's chart of classifying different types of 'giant' images.
- Teacher shows another set of copies of the 'giant' images from the picture books the children have read and uses these to teach the whole class the difference between 'demands' and 'offers', 'close-up' and 'distant' views and 'high angle' and 'low angle' views.

- Collaborative whole class sorting of 'giant' images into these categories conducted by the teacher with student participation.
- Children divided into six groups and each group given a collection of 'giant' images to glue onto a chart under the classifications of 'offer', 'demand', 'close-up' etc.
- Display completed charts on the wall.
- Indicate to the class that, in subsequent lessons, they will be able to make their own illustrated version of *The Selfish Giant*, putting in the images that they think should be in the book in the places they think appropriate.

Week 2: Session 1

- Shared reading of two versions of *The Selfish Giant* explored by the children in the previous week. During shared reading, consolidate children's understanding of the visual grammatical features of the images by labelling 'offers' and 'demands' etc. during the reading, and discussing the effect that these choices of image have on the way you understand the nature of the character and the development of the story.
- Class divided into six groups and each group allocated to one of the following tasks. The groups will rotate three times in this session (i.e. complete three tasks) and rotate three times in the next session – by then completing all tasks. The first task is to explore another two of the illustrated versions of *The Selfish Giant*, chart the types of images and their locations in the text (in the same way this was done in the first lesson using the *Jack and the Beanstalk* stories), and compare the effects of these image choices and their locations across stories. The second task is to choose or construct the images of the giant they want to use in their story and then to scan and/or generate image files stored on a computer. The third task is to read a number of versions of *The Selfish Giant* on the web, noting the absence of images and also paying attention to the format of the screen, font type etc. (Again the audiofile and/or the teacher's recorded reading can be used to support any children who still have difficulty with decoding the story.) The fourth task is to determine the locations in which the group wishes to insert images of the giant into their story, including what kinds of images in which locations. (The fact that some groups do this prior to comparing this aspect of the two published books is not an issue. They can adjust their choices in this task

after they compare the published books if they wish to do so.) The fifth task is to design a front 'cover' or 'banner page' for their online illustrated book. This entails listing Oscar Wilde as the author and themselves as illustrators. Published books will be used as models for this task. The final task involves each group working with the teacher to construct props to use in photographs that will be located on the homepage introducing the children's books posted on the web. This page will say something like 'Grade two class at Sunrise School respond to *The Selfish Giant*'. One set of props will be very small cardboard models of classroom furniture such as a desk and a chair and the digital camera will be used to photograph, using a low angle view, one of the male teachers in the school (or a parent, or student teacher or other adult helper) disguised as a giant. The other props will be a very large cardboard model of a desk and a chair in front of which groups of children in turn will be photographed using a high angle view.

- Conclude with serial reading of *The BFG*.

Week 2: Session 2

- Serial reading of *The BFG*.
- Rotation of group work from previous session.

Week 2: Session 3

- Serial reading of *The BFG*.
- Class divided once again into six groups. Three of these groups will work with the teacher and the other three groups will rotate in this session around the following tasks. The first task is to revisit the 'Fireblade Coffeehouse' site (http://www.hoboes.com/html/FireBlade/Wilde/Giant.html) and review their initial posting of comments about *The Selfish Giant*, adding any new comments in the light of the classroom work they have done. The second task is to design and illustrate an alternative book cover for *The BFG*. The final task is to use the digital camera to take the photographs of themselves beside the large cardboard props they have constructed with the teacher and photographs of the adult helper disguised as a giant, in front of the small props. The adult helper will be able to effect supervision over these groups while the other three groups work with the teacher. These groups will have two main tasks.

One is to actually insert the images into the text of *The Selfish Giant* using a software program that can be used to create a file for eventual display on a website. The second task is to write an account of why they chose particular images for particular locations in the text and what effect they expected these images to have. Both of these tasks will require close teacher support of the students' efforts. In the next session the two sets of three groups of students will 'swap' tasks.

Week 2: Session 4

- Conclude serial reading of *The BFG*.
- Rotation of group work from previous session.

Week 2: Session 5

- Children view their illustrated versions of *The Selfish Giant* on the web. (If the school is not able to host the site on its server, there are a number of sites where the work can be hosted free of charge, such as the geocities site (http://geocities.yahoo.com/ps/learn2/HowItWorks4_Free.html)).
- This session is mainly devoted to rehearsing a presentation of the work done by the class. The presentation could be to a school assembly, a parents' night, or to other class groups in the school. The children will choose one of the following segments to participate in:

    - Introduction of the various 'giant' books that were read including the author and illustrator, and a brief summary. This would include some *Jack and the Beanstalk* stories, *Uncle David*, *The Selfish Giant*, *The BFG* and *Rose Meets Mr Wintergarten*.
    - A Readers Theatre performance of *The Selfish Giant* script provided by Thomas Eastland (http://users.chariot.net.au/~ghost/selgi.htm).
    - A description of the charts of 'giant' images, explaining the different types of images and their effects.
    - A group recorded reading of the *The Selfish Giant* to accompany the website presentations of the illustrated stories.

The Little Prince

The story of *The Little Prince* (de Saint-Exupery, 2000a) was introduced in Chapter 4, where the book was compared with the CD-ROM version including animated illustrations (de Saint-Exupery, 2000b). The story is also available on the web with the original (static) images (http://www.angelfire.com/hi/littleprince/frames.html). In fact, translations in many different languages can be accessed on the web (http://www.geocities.com/athens/rhodes/1916/online.html). One commercial website (http://www.lepetitprince.com/en/) offers a lot of *Little Prince* commodities for sale (such as school cases, pencils, erasers etc.) and also the opportunity to write to *The Little Prince* as well as monthly favourite letters that have in fact been submitted by children from all over the world. This site also advertises a CD-ROM game based on the story – *Learn and Play with The Little Prince*. In this game *The Little Prince* must find the missing part for the astronomer's telescope with the help of the fox and the geographer. The search takes the trio on a journey between their three worlds gathering clues that will help them find the missing part of the telescope. As the telescope is repaired, the astronomer finally shows the Little Prince what cannot be seen by the naked eye: space and its great mysteries. There are many different websites devoted to *The Little Prince*, including testimonials to what has been learned from the story (http://www.geocities.com/razzberryrainstars/littleprince.html) and sites where collectors seek to extend their collections of hundreds of different editions of the book and display images of the various covers they already possess (http://members.lycos.nl/tlp/). The site constructed by Sherry Liu (http://lepetitprince2.tripod.com/) provides the illustrated story in English and French, a story summary, as well as character sketches, notes about symbolism in the story, a message board and links to other *Little Prince* pages.

This outline of ideas for a two-week classroom program for children from about 9 to 12 years of age based around *The Little Prince* focuses on three main areas of work for students. The first is using the web to appreciate the international and inter-generational 'life' of this kind of literary classic and to explore information about the author and the context of composition of the story. The second area of work involves developing an understanding of the 'visual grammar' of images and the role of different grammatical features of images as part of the narrative technique of the book. Some detailed exploration of the images in *The Little Prince*, along the lines of the discussion in Chapter 4, will be complemented by wider reading of other books where the characters have a role in the drawing of the images that appear in the book and hence the construction of the images are actually a part of the plot. Examples of such stories include *Bear Hunt* (Browne, 1982), *Luke's Way of Looking* (Wheatley and Ottley, 1999), *Dreamwalker* (Carmody and Woolman, 2001), and perhaps also Betsy Byars' *The Eighteenth Emergency* (Byars, 1974). The third aspect of the work for students is expanding their literary appreciation of other stories that involve supernatural characters who provide opportunities for characters to gain new insights and understandings. Examples of stories like this include *Skellig* (Almond, 1998), *Megan's Star* (Baillie, 1988) and *The Nargun and the Stars* (Wrightson and Ingpen, 1988).

Week 1: Session 1

- Introduction: show the children a collection of different versions of the story, including the CD-ROM, and indicate that the story also appears on the web. Outline the story briefly as summarized in Chapter 4 of this volume. It may also be useful to indicate that this book is the source of many short quotations that people like to use and there are several that appeal to children such as:

Grown-ups never understand anything by themselves, and it is tiresome for children to be always and forever explaining things to them.

- Read the first two chapters aloud to the class.
- Plan the following six group work activities to begin the next day:
  - *About the author*: Children can access the biography section on the CD-ROM (de Saint-Exupery, 2000b) and a number of websites, such as the Slawek site (http://members.lycos.nl/tlp/antoine.htm), that provide information about the author and how the book came to be written. They can compare this information with the briefer information in the various book versions. Students should select from their findings to produce a chart or 'Powerpoint' slideshow 'about the author'.
  - *International and inter-generational literature*: There are three sections of this task. The first is for the children to visit a selection of websites to note the extraordinary number of languages in which they can access *The Little Prince* story and also to note the virtual communities that exchange ideas about the story as well as work

done by children and portrayed on the websites. As well as the sites listed above, it would be useful for students to visit the dynamic German website, *Der Kleine Prinz* (http://mitglied.lycos.de/kleineprinz/). Notwithstanding the language difference, the children will recognize the range of offerings on this exciting page, including a chat room, postcards, aphorisms, quotes, a picture of planet B612, roses, music and a guest book. The second task is for the children to design and conduct a telephone survey of their own extended family members to find out who has read *The Little Prince*, in what language and under what circumstances in their lives. The final task is to check the school and local libraries to see what version(s) of the story are held in these collections and also to check to see if copies of the story are on sale in local bookshops, as well as checking Amazon.com (where they will also find an advertisement of a DVD of the story (Donen, 2004)). The children should again prepare a chart or electronic report on their telephone, library, bookshop and web surveys.

- *Reading online*: The children should first survey a selection of sites where the illustrated story is available online and note the range of presentation formats, and then select one that appeals to them and read the story. Children should be encouraged to compare their experience of reading a story online with reading in conventional book format. The feedback report from this group will focus on their experience of online reading of the story.
- *Characters drawing their story*: The children first read *Bear Hunt* (Browne, 1982) to observe the ways in which the drawing done by a story character are the drawings in the book and an integral part of the story. They should then explore other books where the images in the book have a somewhat similar narrative role such as *Luke's Way of Looking* (Wheatley and Ottley, 1999) and *Dreamwalker* (Carmody and Woolman, 2001). In reporting their work, groups undertaking this task should select one or two examples of images in each book and explain how characters in the books were involved in their composition and how this is part of the story.
- *Otherworldly characters in other stories*: The children can read *Skellig* or listen to an excerpt online read by the author (http://

www. davidalmond.com/). It is also possible to purchase an audio cassette version of the story read by the author. Here the children might be asked to select a paragraph or two for reading aloud to the class.

- *Image Exploration*: Teacher attention is primarily directed to this group. Here the children are introduced to key elements of the grammar of visual design, focusing on the visual grammar of interactive meanings. Concepts introduced are those of contact – the differences between offers and demands; social distance – long medium and close-up views; power – high, eye-level and low angle views; and involvement – parallel or oblique horizontal angle. One possibility is to use the online story *The Littlest Knight* (Moore, 1994), which was discussed in relation to image analyses in Chapter 5, and then move to a discussion of the grammatical features of the images in the early chapters of *The Little Prince*.

Over the next three days the children will all participate in all of the groupwork tasks. Each of the six groups of children will complete two tasks each day over the next three days. This whole class planning time is important to optimize the students' engagement with and benefit from the tasks which they begin in the next session.

Week 1: Session 2

- Read aloud to the class the next two chapters of *The Little Prince*.
- Children divide into six groups and work on the first two of the groupwork tasks described above.
- Brief review of procedural management of group work tasks.
- Review of what we can now say about the story chapters we have read as a whole class to date. Children will be asked to review these chapters for 'quotable' quotes, which can be added to the class notice board. They can also be asked to comment on the images encountered to date and their role in both the story itself and the questions/issues/lessons raised in the story so far.

Week 1: Session 3

- Introduce children to another book which involves an 'otherworldly' character who enables the main human character to develop deeper insights or understandings. One such story is *Megan's Star* (Baillie, 1988). In this story Megan meets Kel, who has rare powers and knows that Megan has them too. But as they explore their capabilities, Megan realizes she must soon give up all she knows, for there will be no turning back. Read the first two chapters aloud to the children.

- Group work tasks during which all children complete the second set of two tasks for their group.
- Discussion with whole class about ways in which *Megan's Star* and *The Little Prince* are the same and different.

Week 1: Session 4

- Read aloud the next two chapters of *Megan's Star*.
- Group work tasks during which all children complete the final set of two tasks for their group.
- Discussion with whole class about similarities and differences among *Skellig* (which they have all now encountered in group work), *Megan's Star* and *The Little Prince*.

Week 1: Session 5

- Read the next two chapters of *Megan's Star*.
- Provide time for student groups to finalize displays/presentations of the results of their group work.
- Children share displays and presentations of their group work.
- Discussion with the whole class about what has been learned about *The Little Prince*, the narrative techniques of other stories involving 'otherworldly characters', and the role of images in constructing stories.
- Complete reading aloud of *Megan's Star*.

Week 2: Session 1

- Review briefly with the children again the results of their group work tasks as displayed and consolidate particularly the work with the teacher on the grammar of visual design and its use in understanding the role of different types of images in developing the story.
- Using a data projector (or in a computer lab) introduce the class to the CD-ROM story, *Lulu's Enchanted Book* (Victor-Pujebet, n.d.), which was discussed in Chapter 5. In this digital story a princess, named Lulu,

lived in the pages of an enchanted book. One day a flying saucer crashes into her story and she meets the robot pilot, named Mnemo. Mnemo is trying to return to his home planet and his young master, Prince Megalo Polo. Lulu decides to help him and, together, they traverse deserts, jungles, regions of ice, and obstacles peculiar to books: at one point they become lost in the intricacies of the plot (they have rashly attempted to skip a chapter to get water for the spacecraft's cooling system from the chapter on rainforests). These problems are overcome, but the greatest one remains: how can Lulu, a two-dimensional being from a book, make the journey into space – which is by definition three-dimensional? Mnemo realizes he can achieve this by 'extrudomorphosis': a process, he explains, invented by the Italian renaissance painters to allow otherwise flat, paper characters to enter the 3-D world. A happy ending is achieved by playful subversion of physical laws.

- For the remainder of this session and the first part of the next session the class will work in four groups. One will explore all of *Lulu's Enchanted Book* on CD-ROM; a second will similarly explore all of the CD-ROM version of *The Little Prince*; the third group will complete their reading of *Skellig* (using the audio cassettes and/or teacher and/or peer assistance in the case of any students who find reading the book difficult); and the final group will complete their reading of *Dreamwalker*, paying particular attention to the images, and when finished exploring the images on the Dreamwalker websites, such as Cally Steussy's image (http://elfwood.lysator.liu.se/loth/s/t/steussy/nesaka1clrsz.jpg.html).

Week 2: Session 2

- Complete group work tasks from previous session. Early finishers might engage in web searches for reviews of *Lulu's Enchanted Book*, more information about *Skellig* etc.
- The teacher demonstrates the role of image analyses in comparing the interpretive possibilities of two different story versions – in this case the hypertext CD-ROM of *The Little Prince* and the other hard copy and linear online versions the children have read. The focus of the comparison is the geographer chapter in *The Little Prince*, as discussed in detail in Chapter 4.
- The teacher plans with the children the following four group work tasks to be completed over the next two sessions:

- The children select another episode in *The Little Prince* that they will compare, applying the visual analyses used in the teacher demonstration.
- The children will select an episode from *Lulu's Enchanted Book* and undertake an image analysis to indicate the role of the hypertext-activated images in confirming/changing the way the written story is interpreted.
- The children will similarly select an episode from one of the illustrated books read in the previous week such as *Dreamwalker* or *Luke's Way of Looking*, analyse the images and comment on the role of the different types of images and how other image selections might have been used to impact upon the story interpretation.
- Children will select one episode from *Skellig* and design and insert images at points in the text that they determine as appropriate.

Week 2: Sessions 3 and 4

• Each of the four groups in the class completes two of the tasks as outlined above, so that by the end of session four all children have completed all tasks.

Week 2: Session 5

• Reporting by student groups of their four completed tasks.
• Whole class discussion of the different sections of the CD-ROMs that the four groups investigated, the different episodes of the books in which they explored the images, and the different representations for the different episodes of *Skellig* they chose.
• In this session the children next set up a 'poster' presentation of their work and invite (at this point or on a subsequent occasion) other teachers, classes and parents to attend. As visitors circulate around the presentations, students are on duty to explain their work.

Conclusion

The extended programs of classroom learning experiences outlined in this chapter have drawn on classic literary narratives, which have been re-published many times in many different formats and continue to be popularly re-presented now as electronic texts in the age of digital multimedia. Throughout this book, many similar examples of the enduring appeal and social significance of such stories have been noted. The kinds of classroom programs indicated here can also be designed drawing on the new forms of literary narratives composed with and for presentation by digital media in CD-ROM format or on the web. Examples of such emerging forms of digital narratives for children and electronic game narratives as a basis for such classroom work have been discussed in Chapters 5 and 6. In Chapter 3 the hybridization of narrative was explored as a resource for enhancing students' engagement with literary narratives. This hybridization takes account of the increasing tendency of the story world of particular narratives to embrace a complex of dimensions and expressions, which story consumers access through a range of complementary forms including the novel or picture book, the movie and/or DVD, various videogames, website augmentation of the story world in various ways and interactive online sharing of responses and re-creations of the story. If children's school experience of literary narrative is to reflect their potential experience of it as a contemporary cultural phenomenon, it is crucial for teachers to understand and develop the role of e-literature in the English classroom and in literacy learning and teaching. To appreciate the importance of this for schooling, it is salutary to note two recent examples of acknowledged literary narratives as currently experienced in the hybridized multimedia digital context of contemporary storying. The first of these literary texts is the Caldecott medal winning picture book, *The Polar Express* by Chris Van Allsburg (1985). The second is Newbery medal winning novel for children, *Holes* by Louis Sachar (1998).

The story of *The Polar Express* begins one Christmas Eve a long time ago when a young boy lies in bed listening hard for the bells of Santa's sleigh, which a friend has told him do not exist. But what he hears instead is a steam locomotive that pulls in right outside his bedroom window. The conductor invites him to board the Polar Express to the North Pole. The boy and the other children on the train travel to meet Santa and the elves at the North Pole. The boy is selected to receive the first gift of Christmas and he chooses a bell from Santa's sleigh. When the children return on the sleigh the boy realizes he has lost the bell and returns home broken-hearted. But in the morning his sister finds one small present with his name on it. The boy and his sister are enchanted by the sound of the silver bell, but their parents cannot hear any sound coming from it. The boy continues to believe in the spirit of Christmas and can hear the bell even as an adult.

More than 10 years after the book was published, an animated CD-ROM version of the story was produced (Chris Van Allsburg, 1997). In 2004 Warner Brothers have produced a movie version staring Tom Hanks and produced by Robert Zemeckis. The filming technique uses an innovative advanced version of the distinctive 'motion capture' process by which an actor's live performance is digitally captured by computerized cameras and becomes a blueprint for creating virtual characters. Hence Tom Hanks is able to 'play' an eight-year-old boy and the director was able to achieve a representation of the setting consistent with what he believed was a key feature of the book. The Warner Brothers website promoting the film contains a wealth of informational resources and activities about the book, the author and the film (http://polarexpressmovie. warnerbros.com/). The film trailer, of course, can be viewed online, and a great deal of information about the production of the film is provided. There is a biography of the author and online interviews with him as well as a link to the author's website. Teaching suggestions and lesson plans are provided, as well as information about The Polar Express Reading Challenge in North America in which Warner Brothers, Houghton Mifflin Publishers and the National Education Associations combine to offer financial incentives and prizes (such as signed copies of the book and private school screenings of the movie) to encourage young children to read (http://php.warnerbros.com/movies/polarexpress/pages.php?s= challenge). The site also offers downloads of pdf files of paper-based games and activities as well as posters, wallpapers, screensavers and pc video games.

Louis Sachar's book, *Holes* (Sachar, 1998), has also been produced as a movie by Disney starring Sigourney Weaver, Jon Voight, and Shia La Beouf and directed by Andrew Davis. In this story Stanley Yelnats is a young lad who has been falsely accused of stealing sneakers but is nevertheless sent to a boys' juvenile detention centre called Camp Green Lake. This is a desolate place in which Stanley and his peers are required to dig a hole every day, ostensibly because the warden believes the task will be character building for the inmates. However, Stanley and his fellow diggers soon realize that she has other motives and they determine to find out what the real object of digging the holes is. The Disney site promoting the movie includes the trailer, a synopsis, links to the author's website, downloadable lesson plans, and many other informational resources about the movie as well as an online game related to the story. The author's website provides a wealth of information about *Holes* and his other books (http://www. louissachar.com/), but the book has also spawned many other websites. One such site provides a list of multiple links to other *Holes* sites (http://www.nvo.com/ecnewletter/holesbylouissachar/) including lesson plans, book discussion guides and a 'thinkquest' (http://library.thinkquest.org/J0113061/).

The strength of appeal and value of literary narratives for children such as these has clearly assured them of a central role in the emerging digital multimedia world through which contemporary story is experienced. More than 10 years ago Margaret Mackey indicated that:

> To talk about children's literature, in the normal restricted sense of children's novels, poems and picture-books, is to ignore the multi-media expertise of our children.
>
> (Mackey 1994: 17)

The potential of the expanded digital context of story worlds as a resource for encouraging sustained reading of literary narratives among young people (Mackey, 2001) needs to take account of the impact of ICT on the textual practices surrounding literary texts and, indeed, on the character of literary narratives themselves (Locke and Andrews, 2004), changing the very nature of what we understand to be narratives (Hunt, 2000). Locke and Andrews cite Donald Leu (2000) in suggesting that in responding to the imperative for research into the impact of ICT on how students are working with traditional and new literary forms, it may well be that 'teachers themselves, exploring in their own classrooms hunches and intuitions about the implications for their teaching' can 'provide the strongest lead as to how the future research agenda should be formulated' (2004: 148). It is hoped that this book will resource and encourage teachers to undertake this kind of exploration.

References

Ahlberg, J. and Ahlberg, A. (1986) *The Jolly Postman, or, Other People's Letters*. London: Heinemann.

Almond, D. (1998) *Skellig*. London: Hodder.

Alvermann, D. (ed.) (2004) *Adolescents and Literacies in a Digital World*. New York: Peter Lang.

Andrews, R. (2004a) 'Where next in research on ICT and literacies?', *Literacy Learning: The Middle Years*, 12(1), 58–67.

Andrews, R. (ed.) (2004b) *The Impact of ICT on Literacy Education*. London and New York: RoutledgeFalmer.

Ashby, R. (2001) *Shrek*. New York: ipicturebooks.com.

Astorga, C. (1999) 'The text-image interaction and second language learning', *Australian Journal of Language and Literacy*, 22(3), 212–33.

Austin, H. (1993) 'Verbal art in children's literature: an application of linguistic theory to the classroom', *English in Australia*, 103, 63–75.

Baillie, A. (1988) *Megan's Star*. Melbourne: Nelson.

Bawden, N. (1981) *William Tell*. London: Cape.

Bearne, E. (2000) 'Past perfect and future conditional: the challenge of new texts', in G. Hodges, M. Drummond and M. Styles (eds) *Tales, Tellers and Texts* (pp. 145–56) London: Continuum.

Brown, M. (1999) *Arthur's First Sleepover*. Sydney: Scholastic Australia.

Browne, A. (1982) *Bear Hunt*. London: Scholastic.

Browne, A. (1994) *Zoo*. London: Random House.

Browne, A. (1986) *Piggybook*. London: Julia MacRae.

Buff, C. (2001) *The Apple and the Arrow*. Boston: Houghton-Mifflin.

Burnett, F.H. (1992) *The Secret Garden*. London: Sainsbury Walker.

Burningham, J. (1984) *Grandpa*. London: Penguin/Puffin.

Byars, B. (1974) *The Eighteenth Emergency*. London: The Bodley Head.

Callow, J. and Zammit, K. (2002) 'Visual literacy: from picture books to electronic texts', in M. Monteith (ed.) *Teaching Primary Literacy with ICT* (pp. 188–201) Buckingham: Open University Press.

Cannon, J. (1996) *Stellaluna*. San Francisco: LivingBooks/Random House/Broderbund.

Carmody, I. (1987) *Obernewtyn*. Ringwood, Victoria: Puffin.

Carmody, I. (1990) *The Farseekers*. Ringwood, Victoria: Viking.

Carmody, I. (1992) *Scatterlings*. Ringwood, Victoria: Puffin.

Carmody, I. (1993) *The Gathering*. Ringwood, Victoria: Penguin.

Carmody, I. (1996) *Ashling*. Ringwood, Victoria: Puffin.

Carmody, I. and Woolman, S. (2001) *Dreamwalker*. Melbourne: Lothian.

Carroll, L. (2000) *Alice's Adventures in Wonderland* (CD-ROM) Brighton: Joriko Interactive.

Caswell, B. and Ottley, M. (2003) *Hyram and B*. Sydney: Hodder Headline.

Chambers, A. (1983) *The Present Takers*. London: Bodley Head.

Chandler-Olcott, K. and Mahar, D. (2003) ' "Tech-saviness" meets multiliteracies: exploring adolescent girls technology-related literacy practices', *Reading Research Quarterly*, 38(10), 356–85.

Cope, B. and Kalantzis, M. (eds) (2000) *Multiliteracies: Literacy Learning and the Design of Social Futures*. Melbourne: Macmillan.

Cleary, B. (1976) *Ramona the Pest*. Harmondsworth: Puffin.

Cleary, B. (1978) *Ramona the Brave*. Harmondsworth: Puffin.

Cleary, B. (1981) *Ramona and her Father*. Harmondsworth: Puffin.

Cleary, B. (1982) *Ramona and her Mother*. Harmondsworth: Puffin.

Cleary, B. (1984) *Ramona Quimby, Age 8*. Harmondsworth: Puffin.

Cleary, B. (1986) *Ramona forever*. Harmondsworth: Puffin.

Crane, S. (1974) *The Red Badge of Courage*. New York: Scholastic.

Dahl, R. (1988) *Matilda*. London: Jonathan Cape.

Dahl, R. and Blake, Q. (1982) *The BFG*. London: J. Cape.

Davidson, J. (2000) 'Young movie makers', *Practically Primary*, 5(2), 8–9.

de Saint-Exupery, A. (2000a) *The Little Prince*. London: Penguin.

de Saint-Exupery, A. (2000b) *The Little Prince* (CD-ROM) Tivola.

Disney (1998) *Mulan Animated Storybook* (CD-ROM) Burbank, CA: Disney Interactive.

Donen, S. (Writer) (2004) *The Little Prince*. USA and Canada: Paramount Home Video.

Doonan, J. (1993) *Looking at Pictures in Picture Books*. Stroud: Thimble Press.

Doyle, A. (1982) *The Hound of the Baskervilles*. Harmondsworth: Puffin.

Dresang, E. (1999) *Radical Change: Books for Youth in a Digital Age*. New York: Wilson.

Dresang, E. and McClelland, K. (1999) 'Radical change: digital age literature and learning', *Theory into Practice*, 38(3), 160–7.

Durrant, C. and Hargreaves, S. (1995) 'Literacy online: the use of computers in the secondary classroom', *English in Australia*, 111, 37–48.

Early, M. (1991) *William Tell*. Sydney: Walter McVitty.

Education Queensland (1995) *English 1–10 Syllabus: A Guide to Analysing Texts*. Brisbane: Queensland Government Printing Office.

Electronic Arts (2002) *Harry Potter and the Chamber of Secrets Adventure Game* (CD-ROM) Warner Brothers.

Electronic Arts (2004a) *Harry Potter and the Prisoner of Azkaban* (CD-ROM) Warner Brothers.

Electronic Arts (2004b) *Harry Potter: Quidditch World Cup* (CD-ROM) Warner Brothers.

Electronic Arts (2004c) *The Lord of the Rings: The Return of the King* (CD-ROM) Warner Brothers.

Fisher, L. (1996) *William Tell*. New York: Farrar Strauss and Giroux.

Freeman, D. (1975) *Corduroy*. Harmondsworth: Puffin.

Gallimard. (2000) *The Little Prince*. Milton Keynes: Tivola/Editions Gallimard.

Gee, J.P. (2003) *What Computer Games have to Teach us about Learning and Literacy*. New York: Palgrave Macmillan.

Gleeson, L. and Greder, A. (1992) *Uncle David*. Sydney: Ashton Scholastic.

Golding, W. (1963) *The Lord of the Flies*. Harmondsworth: Penguin.

Goodman, S. and Graddol, D. (1996) *Redesigning English: New Texts, New Identities*. London: Routledge.

Graham, B. (1992) *Rose Meets Mr Wintergarten*. Ringwood, Victoria: Viking.

Guyer, C. and Joyce, M. (2000) *Lasting Image*. Retrieved 11/8/2004, from http://www. eastgate.com/LastingImage/Welcome.html.

Halliday, M.A.K. (1973) *Explorations in the Functions of Language*. London: Arnold.

Halliday, M.A.K. (1978) *Language as a Social Semiotic: The Social Interpretation of Language and Meaning*. London: Edward Arnold.

Halliday, M.A.K. (1994) *An Introduction to Functional Grammar* (2nd edn) London: Edward Arnold.

Halliday, M.A.K. and Hasan, R. (1985) *Language, Context and Text: Aspects of Language in a Social-Semiotic Perspective*. Geelong: Deakin University Press.

Halliday, M.A.K. and Matthiessen, C. (2001) *An Introduction to Functional Grammar* (3rd edn) London: Arnold.

Hasan, R. (1985) *Linguistics, Language and Verbal Art*. Geelong: Deakin University Press.

Hasan, R. (1995) 'The conception of context in text', in P. Fries and M. Gregory (eds) *Discourse in Society: Systemic Functional Perspectives* (pp. 183–283) Norwood, NJ: Ablex.

Hathorn. (1999) *The Wishing Cupboard*. Retrieved 11/8/2004 from http://www. libbyhathorn.com/lh/Wishing/Default.htm.

Hodges, G., Drummond, M. and Styles, M. (eds) (2000) *Tales, Tellers and Texts*. London: Continuum.

Houghton Mifflin. (1995) *The Polar Express* (CD-ROM) Boston, MA: Houghton Mifflin Interactive.

Howard, P. (2001) *XYLO*. Retrieved 11/8/2004, from http://www.wordcircuits.com/gallery/xylo/index.html.

Humphrey, S. (1996) *Exploring Literacy in School Geography*. Sydney: Metropolitan East Disadvantaged Schools Program, New South Wales Department of School Education.

Hunt, P. (2000) 'Futures for children's literature: evolution or radical break', *Cambridge Journal of Education*, 30(1), 111–19.

InnerWorkings. (1997) The Jolly Post Office. Pyrmont, NSW: Roadshow Interactive.

James, R. (1999) 'Navigating CD-ROMs: an exploration of children reading interactive narratives', *Children's Literature in Education*, 30(1), 47–63.

Jennings, P. and Gleitzman, M. (1998) *Wicked! All Six Books in One*. Ringwood, Victoria: Puffin.

Jewitt, C. (2002) 'The move from page to screen: the multimodal reshaping of school English', *Visual Communication*, 1(2), 171–96.

Jiang, W. (1997) *The Legend of Mu Lan: A Herione of Ancient China*. Monterey, CA: Victory Press.

Joyce, W. (1985) *George Shrinks*. New York: Harper Collins.

Joyce, W. (c.1994) *George Shrinks* (CD-ROM) New York: Harper Collins Interactive.

King, C. (1963) *Stig of the Dump*. Harmondsworth: Viking Kestrel.

Knowles, M. and Malmkjaer, K. (1996) *Language and Control in Children's Literature*. London: Routledge.

Kress, G. and van Leeuwen, T. (1990) *Reading Images*. Geelong: Deakin University Press.

Kress, G. and van Leeuwen, T. (1996) *Reading Images: A Grammar of Visual Design*. London: Routledge.

Lankshear, C. and Knobel, M. (2003) *New Literacies: Changing Knowledge and Classroom Learning*. Buckingham/Philadelphia: Open University Press.

Lankshear, C., Snyder, I. and Green, B. (2000) *Teachers and Technoliteracy*. Sydney: Allen and Unwin.

Larsen, D. (1999) *Stained Word Window*. Retrieved 11/8/2004, from http:// wordcircuits. com/gallery/stained/index.html.

Ledgerwood, M. (1999) 'Multimedia literature, "exploratory games" and their hypertextuality', in S. Inkinen (ed.) *Mediapolis: Aspects of Texts, Hypertexts and Multimedial Communication* (pp. 44–56) Berlin and New York: de Gruyter.

Lee, J.M. (1995) *The Song of Mulan*. Arden, NC: Front Street.

Left Handed Creations (1994–2004) *Banpf*. Retrieved 24/8/2004, from http:// www. banph.com/.

Lemke, J. (1998a) 'Metamedia literacy: transforming meanings and media', in D. Reinking, M. McKenna, L. Labbo and R. Kieffer (eds) *Handbook of Literacy and Technology: Transformations in a Post-typographic World* (pp. 283–302) Mahwah, NJ: Erlbaum.

Lemke, J. (1998b) 'Multiplying meaning: visual and verbal semiotics in scientific text', in J.R. Martin and R. Veel (eds) *Reading Science: Critical and Functional Perspectives on Discourses of Science* (pp. 87–113) London: Routledge.

Lewis, D. (2001) *Reading Contemporary Picturebooks*. London: RoutledgeFalmer.

Lewis, K. and Moore, C. (2003) *Wolstencroft The Bear*. Retrieved 2/8/2004, from http://www.magickeys.com/books/wolstencroft/index.html.

Locke, T. and Andrews, R. (2004) 'ICT and literature: a Faustian compact?', in R. Andrews (ed.) *The impact of ICT on Literacy Education* (pp. 124–52) London and New York: RoutledgeFalmer.

Mackey, M. (1994) 'The new basics: learning to read in a multimedia world', *English in Education*, 28(1), 9–19.

Mackey, M. (1999) 'Playing the phase space', *Signal*, 88, 16–33.

Mackey, M. (2001). 'The survival of engaged reading in the internet age: new media, old media, and the book', *Children's Literature in Education*, 32(3), 167–89.

Mackey, M. (2002) *Literacies Across Media: Playing the Text*. London and New York: RoutledgeFalmer.

Malory, T. (1975) *King Arthur and his Knights: Selected Tales*. London and New York: Oxford University Press.

Marsden, J. and Tan, S. (1998) *The Rabbits*. Melbourne: Lothian.

Martin, J.R. (1991) 'Intrinsic functionality: implications for contextual theory', *Social Semiotics*, 1(1), 99–162.

Martin, J.R. (1992) *English Text: System and Structure*. Amsterdam: Benjamins.

Martin, J.R. and Rose, D. (2003) *Working with Discourse: Meaning Beyond the Clause* (1st edn, Vol. 1) London/New York: Continuum.

Mattel Media (2000) *Arthur's Absolutely Fun Day!*. Mattel Media.

Matthiessen, C. (1995) *Lexicogrammatical Cartography: English Systems*. Tokyo: International Language Sciences.

Matus, M. (2002) *The Inner Circle – 1st Book of the Relic Triangle*. Retrieved 3/8/2004, from http://www.relictriangle.com/.

McCloud, S. (1994) *Understanding Comics: The Invisible Art*. New York: Harper Collins.

McCloud, S. (2000) *Reinventing Comics*. New York: HarperCollins.

Merino, R. (2004) *The Farm Animals*. Retrieved 2/8/2004, from http://www.magickeys. com/books/farm/page1.html.

Miller, L. (2000) 'Matilda: Finland's telematic literature project', *Reading Online*, 4 (3) Available: http://www.readingonline.org/international/inter_indsx.asp?HREF=/international/miller2/index.html.

Miller, L. and Olsen, J. (1998) 'Literacy research oriented toward features of technology and classrooms', in D. Reinking, M. McKenna, L. Labbo and R. Kieffer (eds) *Handbook of Literacy and Technology: Transformations in a Post-typographic World* (pp. 343–60) Mahwah, NJ: Erlbaum.

Misson, R. (1998) 'Telling tales out of school', in F. Christie and R. Misson (eds) *Literacy and Schooling*. London: Routledge.

Moore, C. (1994) *The Littlest Knight*. Retrieved 3/8/2004, from http://www.magickeys.com/books/lk/index.htm#start.

Moore, C. (2001) *Wind Song*. Retrieved 19/8/2004, http://www.magickeys.com/books/windsong/index.html.

Moore, C. and Paulhamus, J. (2003) *Second Thoughts*. Retrieved 24/8/2004, from http://www.magickeys.com/books/alien/index.html.

Morgan, W. (2002) 'Heterotropes: learning the rhetoric of hyperlinks', *Education, Communication and Information*, 2(2/3), 215–33.

Morgan, W. and Andrews, R. (1999) 'City of text? Metaphors for hypertext in literary education', *Changing English*, 6(1), 81–92.

Munsch, R. (1994) *The Paper Bag Princess* (CD-ROM) Buffalo, NY: Discis.

Murray, J. (1998) *Hamlet on the Holodeck: The Future of Narrative in Cyberspace*. Cambridge, MA: MIT Press.

Nan Zhang, S. (1998) *The Ballad of Mulan*. Union City, CA: Pan Asian Publications.

New London Group (1996) 'A pedagogy of multiliteracies: designing social futures', *Harvard Educational Review*, 66(1), 60–91.

New London Group (2000) 'A pedagogy of multiliteracies: designing social futures', in B. Cope and M. Kalantzis (eds) *Multiliteracies: Literacy Learning and the Design of Social Futures*. Melbourne: Macmillan.

Newman, M. (2004) *Cyberlit*. Lanham, MD and Oxford: Scarecrow.

New South Wales Board of Studies (1998) *English K-6 Syllabus and Support Documents*. Sydney: New South Wales Government.

Ng, T. (2004) *Tiger Son*. Retrieved 9/11/2004, from http://www.magickeys.com/books/tigerson/index.html.

Nodelman, P. (1988) *Words About Pictures: The Narrative Art of Children's Picture Books*. Athens: University of Georgia Press.

O'Brien, R. (1998) *Z for Zachariah*. London: Puffin/Penguin.

O'Toole, M. (1994) *The Language of Displayed Art*. London: Leicester University Press.

Pavic, M. (2003) *The Glass Snail*. Retrieved 4/8/2004, from http://wordcircuits.com/gallery/glasssnail/.

Ponzio, R. (n.d.) *St Mary's Avenue*. Retrieved 3/8/2004, from http://www.sundogstories. net/stmary/index.html.

Ponzio, R. and Labate, J. (n.d.) *Flying Angels*. Retrieved 24/8/2004, from http://www. sundogstories.net/flying/index.html.

Pratchett, T. and Kidby, P. (2001) *The Last Hero*. New York: Discworld/Harper Collins.

Pullman, P. (1995) *Northern Lights*. London: Scholastic.

Pullman, P. (1997) *The Subtle Knife*. London: Scholastic.

Pullman, P. (2000) *The Amber Sypglass*. New York: Alfred Knopf.

Quinn, M. (2004) 'Talking with Jess: looking at how metalanguage assisted explanation writing in the middle years', *Australian Journal of Language and Literacy*, 27(3), 245-261.

Rassool, N. (1999) *Literacy for Sustainable Development in the Age of Information*. Clevedon: Multilingual Matters.

Ravelli, L. (2000) 'Getting started with functional analysis of texts', in L. Unsworth (ed.) *Researching Language in Schools and Communities: Functional Linguistic Perspectives* (pp. 27–64) London: Cassell.

Richards, C. (2001) 'Hypermedia, internet communication, and the challenge of redefining literacy in the electronic age', *Language Learning and Technology*, 4(2), 59–77.

Rock, J. (n.d.) *The Vasalisa Project*. Retrieved 5/8/2004, from http://www.rockingchair.org/.

Rowling, J.K. (1997) *Harry Potter and the Philosopher's Stone*. London: Bloomsbury.

Russell, G. (2000) 'Print-based and visual discourses in schools: implications for pedagogy', *Discourse: Studies in the Cultural Politics of Education*, 21(2), 205-17.

Sachar, L. (1998) *Holes*. New York: Farrar, Straus and Giroux.

San Souci, R. (2000) *FA Mulan: The Story of a Warrior Woman*. New York: Hyperion.

Schleppegrell, M., Achugar, M. and Oteíza, T. (2004) 'The grammar of history: enhancing content-based instruction through a functional focus on language', *TESOL Quarterly*, 38(1), 67–93.

Scieszka, J. and Smith, L.I. (1992) *The Stinky Cheese Man and Other Fairly Stupid Tales*. New York: Viking.

Sefton-Green, J. (2001) 'ICT, the home and digital cultures', in C. Durrant and C. Beavis (eds) *P(ICT)URES of English: Teachers, Learners and Technology* (pp. 162–74) Kent Town, South Australia: Wakefield Press/Australian Association for the Teaching of English.

Sefton-Green, J. and Buckingham, D. (1998) 'Digital visions: children's "creative" uses of multimedia technologies', in J. Sefton-Green (ed.) *Digital Diversions: Youth Culture in an Age of Multimedia*. London: University College London Press.

Sendak, M. (1962) *Where the Wild Things Are*. London: The Bodley Head.

Simpson, A. (2004) 'Book raps as online multimodal communication: towards a model of interactive pedagogy', *International Journal of Learning*, 10, 2705–14.

Small, T. (1991) *The Legend of William Tell*. New York: Bantam/Doubleday.

SQUARE-ENIX. (2002) Final Fantasy X: SQUARE ENIX USA INC.

Steen, B. (2000a) *Jack and the Beanstalk*. Retrieved 2/8/2004, from http://www. tumblebooks.com/syndication/chickadee/indexwf.html.

Steen, B. (2000b) *Old Mother Hubbard*. Retrieved 2/8/2004, from http:// www.tumblebooks. com/syndication/chickadee/indexwf.html.

Steig, W. (1990) *Shrek*. New York: Michael Di Capua Books.

Steinbeck, J. (1937) *Of Mice and Men*. London: Penguin.

Steinbeck Series (1996) *Of Mice and Men*. New York: Penguin Electronics.

Stephens, J. (1994) 'Signifying strategies and closed texts in Australian children's literature', *Australian Review of Applied Linguistics*, 17(2), 131–46.

Stephens, J. (2000) 'Modality and space in picture book art: Allen Say's *Emma's Rug*', *CREArTA*, 1(1), 45–59.

Street (1999) *The Geode Space*. Retrieved 21/9/004, from http:// www.mindspring.com/~roadrunner1/stories/geode_space/storm/storm1.html.

Strickland, S. (1999) *The Ballad of Sand and Harry Soot*. Retrieved 12/8/2004, from http://wordcircuits.com/gallery/sandsoot/index.html.

Styles, M. and Arizpe, E. (2001) ' "A gorilla with Grandpa's eyes": how children interpret visual texts – a case study of Anthony Browne's "Zoo" ', *Children's Literature in Education*, 32(4), 261–81.

Swift, J. (1972) *Gulliver's Travels*. London: Longman.

Swigart, R. (2002) *About Time*. Retrieved 4/8/2004, from http://wordcircuits.com/ gallery/abouttime/.

Thomas, A. (2000) 'Textual constructions of children's online identity', *CyberPsychology and Behaviour*, 3(4), 665–72.

Thomas, A. (2001) 'The cyber child', *disClosure*, 10, 143–75.

Thomas, A. (2004a) 'Children online: learning in a virtual community of practice', *e-learning*, 2(1), 27–38.

Thomas, A. (2004b) 'Digital literacies of the cybergirl', *e-learning*, 1(3), 358–82.

Tivola (2001) *Snow White and the Seven Hansels*. Game. New York.

Tolkien, J.R.R. and Lee, A. (2002) *The Fellowship of the Ring: Being the First Part of Lord of the Rings*. London: Harper Collins.

Topologika (2000/1) *Stig of the Dump*. Penryn, Cornwall.

Ubisoft (in press) *Secret of the Silver Earring*. London: Ubisoft Multimedia.

Ubisoft. (n.d.) *Payuta and the Ice God*. London: Ubisoft Multimedia.

Unsworth, L. (1993) 'Managing the language program: children's literature in the primary school', in L. Unsworth (ed.) *Literacy Learning and Teaching: Language as Social Practice in the Primary School*. Melbourne: Macmillan.

Unsworth, L. (1999) 'Explaining school science in book and CD ROM formats: using semiotic analyses to compare the textual construction of knowledge', *International Journal of Instructional Media*, 26(2), 159–79.

Unsworth, L. (2001) *Teaching Multiliteracies Across the Curriculum: Changing Contexts of Text and Image in Classroom Practice*. Buckingham: Open University Press.

Unsworth, L. (2002) 'Reading grammatically: exploring the constructedness of literary texts', *L1 Educational Studies of Language and Literature*, 2(2), 212–40.

Unsworth, L. (2003) 'Re-framing research and practice relating to CD ROM narratives in classroom literacy learning: addressing "radical change" in digital age literature for children', *Issues in Educational Research*, 13(2), 55–70.

Unsworth, L. (in press) 'Multiliteracies and multimodal text analysis in classroom work with children's literature', in T. Royce and W. Bowcher (eds) *Perspectives on the Analysis of Multimodal Discourse*. Mahwah, NJ: Erlbaum.

Unsworth, L. and Wheeler, J. (2002) 'Re-valuing the role of images in reviewing picture books', *Reading: Language and Literacy*, 36(2), 68–74.

Van Allsburg, C. (1985) *The Polar Express*. London: Andersen.

Van Allsburg, C. (1997) *The Polar Express* (CD-ROM) Somerville, MA: Houghton Mifflin Interactive.

van Leeuwen, T. and Humphrey, S. (1996) 'On learning to look through a geographer's eyes', in R. Hasan and G. Williams (eds) *Literacy in Society* (pp. 29–49) London: Addison Wesley Longman.

Veel, R. (1998) 'The greening of school science: ecogenesis in secondary classrooms', in J.R. Martin and R. Veel (eds) *Reading Science: Functional and Critical Perspectives on the Discourses of Science*. London: Routledge.

Victor-Pujebet, V. (n.d.) *Lulu's Enchanted Book*. Hove: Wayland.

Wells, H.G. (1986) *The War of the Worlds*. New York: New American Library.

Wheatley, N. and Ottley, M. (1999) *Luke's Way of Looking*. Sydney: Hodder Headline.

Wilde, O. (2001) *The Selfish Giant*. Lanham, MD: Derrydale.

Wilde, O., Foreman, M. and Wright, F. (1986 (1888)) *The Selfish Giant*. Harmondsworth: Picture Puffin.

Wilde, O. and Gallagher, S. (1995) *The Selfish Giant*. New York: Putnam.

Wilde, O. and Neale, S. (1994) *The Selfish Giant*. Avonmouth, Bristol: Paragon.

Williams, G. (1993) 'Using systemic grammar in teaching young learners: an introduction', in L. Unsworth (ed.) *Literacy Learning and Teaching: Language as Social Practice in the Primary School*. Melbourne: Macmillan.

Williams, G. (1998) 'Children entering literate worlds', in F. Christie and R. Misson (eds) *Literacy in Schooling* (pp. 18–46) London: Routledge.

Williams, G. (2000) 'Children's literature, children and uses of language description', in L. Unsworth (ed.) *Researching Language in Schools and Communities: A Functional Linguistic Perspective*. London: Cassell.

Williams, M. (2002) 'Writing scary stories in the middle school', *PEN: Primary English Notes*, 136.

Wizard AnimatedClassics (n.d.) *Mulan*. Hong Kong: Rainbow Entertainment.

Wodehouse, P.G. (1904) *William Tell told Again*. London: Adam and Charles Black.

Wrightson, P., and Ingpen, R. (1988) *The Nargun and the Stars*. Hawthorn: Hutchinson Australia.

Zancanella, D., Hall, L. and Pence, P. (2000) 'Computer games as literature', in A. Goodwyn (ed.) *English in the Digital Age* (pp. 87–102) London: Cassell.

Zander, P. (Writer) (1992) *Selfish Giant*, in F.H. Entertainment (Producer) USA.

Zervos, K. (2001) *Childhood in Richmond*. Retrieved 11/8/2004, from http:// wordcircuits. com/gallery/childhood/index.html.

Index

About Time (Swigart) 26, 102–4
accessing, in story-focused games
    127–8
'action lines' 16–17
adjunct puzzles/tasks 125–7
adjunct-composition contexts 51–4
adverbs *see* Circumstances
Alibris 41
Alice's Adventures in Wonderland (Carroll)
    49–50; game narrative of 128–9, 132,
    133
Allsburg, C., *Polar Express* 154–5
Almond, D.: *Skellig* 39, 47, 149–50, 151,
    152, 153; website 41, 149–50
Amazon.com 41
Andrews, R. 156
animations: in CD-ROM versions of
    literary texts 59–60, 63–5(*Table 4.1*),
    65–7, 155; in e-comics 117; in
    e-poetry 114
Ant Lion, The (Wright) 31
appreciation, of reading 43–4; in work
    programs 148
Arcane game 124, 130
Arthur's Absolutely Fun Day game (Mattell
    Media) 133–4
assembling, in story-focused games
    128–9
authors: biographies in work programs
    148; 'fan' sites for 43–4; websites of
    38–41, 46, 55, 98, 100, 156
avatars 52

Baillie, A., *Megan's Star* 151
Ballad of Mulan (Zhang) 41, 69–70
Ballad of Sand and Harry Soot, The
    (Strickland) 115
Banpf (Left Handed Creations) 23,
    98–100

BBC, *Spywatch* 47, 123–4
Bear Hunt (Browne) 149
Bearne, K. 6
BFG, The (Dahl) 139, 141, 142, 143, 145,
    146
blogging (web logging) 50–1
Book Chat site 48
book raps 47–9
book-based literary narratives 2–3, 37;
    adjunct-composition 51–4;
    appreciation 43–4; compared with
    electronic narratives 2–5(*Figure 1.1*),
    57–8, 94; interpretation 45–51;
    invitation 41–3; literature on the web
    54–5; story composition 37–40, *see also*
    CD-ROM; digitally originated
    literary text; electronically
    re-contextualized literary texts
books (printed): and electronic formats
    compared 58–85; story-focused games
    based on 124–5, 132–3; used in work
    programs 138, 147, 149
booksellers, online access to 41, 43
Bouchet, B., *French Letters* 42
Brown, M., Arthur stories 134
Browne, A. 18; *Bear Hunt* 149; *Piggybook*
    18, 28–9
Bunyips (website) 23
Burningham J., *Grandpa* 18

Cannon, J. *Stellaluna* 58, 65–8
captions 84
Carmody, I.: *The Dreamwalker* (Carmody
    and Woolman) 45, 134, 148, 149, 152,
    153; 'fan' sites 43–4, 52
Carol Hurst site 54
Carroll, L., *Alice's Adventures in
    Wonderland* 49–50, 128–9; game
    narrative of 128–9, 132, 133

CD-ROM: animations in versions of
literary texts 59–60, 63–5(*Table 4.1*),
65–7, 155; illustrated
re-contextualizations of classic stories
71–85; stories and images compared
with original literary texts 3–4, 58–71,
151–2; 'talking books' 106; versions of
The Little Prince 147, 148, 149
Chambers, A.: *The Present Takers* 39–40;
website 39
character role, in electronic games 121–2,
125, 131, 133
character sketches, in *Banpf* 99
characters: drawing own stories 149;
otherworldly 149–50, 150–1
'chatrooms' 102
Childhood in Richmond (Zervos) 113–14
children *see* students
Children's Storybooks Online site 90,
91, 94, 95, 96
'choose your own adventure' 100, 105
Circumstances (adverbs) 27, 28
Clarke, B. 123
classic stories, electronically re-
contextualized versions of 71–85
classroom work: groupings for 10; online
lesson plans 45–6; programs of work
for 9–11, 137–8; use of electronic
game narratives in 129–35; use of
hypermedia narratives 105–6; and
Wolstencroft 93, 94, *see also* programs of
work
Cleary, B., website 41, 45
Clues game 124
colour, use in images 20
communication, online: for discussion
46–51; for learning activities 45–6
composition 8, 9, 37–40; adjunct-
composition contexts 51–4
compositional meanings 15
conceptual images 18
Crane, S.: *The Red Badge of Courage* 40;
The Veteran 40
'cropped' images 85
'crosshairs' 114

Dadd, P., images for *William Tell Told
Again* 73–5(*Figures 4.3 and 4.5*)
Dahl, Roald, *The BFG* 139, 141, 142,
143, 145, 146
Dalton elementary school (New York)
49–50, 139
Davidson, J. 133

Davies, I.D. 138
Dead of Night 100
'demand' images 19, 97–8
dependent clause 83(*Figure 4.6*)
Desktop Author 3 software 50
dialogue techniques 69–70, 79–84(*Tables
4.6–4.7*)
Dickens, C., *Christmas Carol, A* 85
digital poetry 88, 114–15
digitally originated literary text 4,
87–90(*Figure 5.1*); digital poetry
114–15; 'e-comics' 116–17;
e-narratives and interactive story
contexts 98–105; e-stories for early
readers 90–4; and electronic games
activities 131–5; hybridization of
154–6; hypermedia narratives 105–14;
hypertext narratives 102–5; linear
e-narratives 94–8
Discover School site 40, 45
Disney, *Mulan Animated Storybook*
(CD-ROM) 30, 68–9, 70–1, 125, 129
Disney website 155–6
Doyle, A.C., Sherlock Holmes stories
134
Dragonroot Cantina (online 'chatroom')
102
Dreamwalker, The (Carmody and
Woolman) 45, 134, 148, 149, 152, 153
*Dreamwalker: Roleplaying in the Land of
Dreams* website 134
Dresang, E. 6, 36
dynamic e-poetry 114

'e-comics' 88, 116–17
e-narratives, and interactive story
contexts 4, 87–8, 89(*Figure 5.1*),
98–105
E-pal project 48
e-poetry 88, 114
e-stories, for early readers 87, 89(*Figure
5.1*), 90–4
Early, M., *William Tell* 17, 18, 26, 30, 32;
CD-ROM and book versions
compared 72–84(*Figures 4.4–5.5,
Tables 4.5–4.7*)
Eastgate website 113
Eastland, T. 139, 146
Edinger, M. 49–50
Electronic Arts: *Harry Potter and the
Chamber of Secrets* game 125, 133; *The
Lord of the Rings: The Return of the King*
game 125

electronic games narratives 4, 41–2, 68–9;
and development of literacy and
literary understanding 129–35;
frameworks for 122–9(*Figure 6.1*);
Liu, S. 147; video games as
'embodied' stories 120–2
Electronic Literature and Literacies in
International Education (ELLIE)
website 47
electronically augmented literary texts *see*
book-based literary narratives
electronically re-contextualized literary
texts 3, 57–8; classic stories online
71–84; multimedia CD-ROM
reconstructions 58–71, *see also* book-
based literary narratives; digitally
originated literary text
ELLIE (Electronic Literature and
Literacies in International Education)
website 47
email discussion groups 46, 48–9
enacting, in story-focused games 128–9,
132–3
Enchanted Forest website 44
Europe of Tales, A website 18
de Saint-Exupery, A.: websites 147, 148,
see also Little Prince

'fan' sites 4, 5, 43–4
Farm Animals, The (Merino) 90
feeling verbs, visual grammars for 27, 29,
91, 92(*Figure 5.2*)
Final Fantasy X game (SQUARE-ENIX)
134–5
Finland, Netlibris site 46–7
Fireblade Coffeehouse site 139, 142, 145
Flying Angels (Ponzio and Labate) 94
Foreman, M. 138
Fox Taming Game 126–7
frameworks *see* electronic games
narratives; interpretive frameworks;
organizational frameworks; pedagogic
frameworks
framing 22–3
French Letters (Bouchet) 42
functional grammar *see* grammatical
readings

Gallagher, S. 138
game-focused stories 122–4
games *see* electronic games narratives
Gee, J. 119, 120, 121–2
geocities website 146

Geode Space, The game (Street) 127, 132
George Shrinks (Joyce) 58–60, 85, 134
girls, websites for 46
Glass Snail (Pavic) 104
Gleeson, L., *Uncle David* 140, 141
Gleitzman, M.: website of 41, 42, 46;
Wicked stories 42
Golding, William, 'fan' site 44
Graham, B., *Rose Meets Mr Wintergarten*
140, 142–3
grammar, visual and verbal *see*
metalanguage; point of view;
Processes
'grammar of visual design' 13, 15, 16–26
grammatical readings 7, 27–36
Grandpa (Burningham) 18
graphic novels (e-comics) 88, 116–17
Greder, A. (illustrator), *Uncle David*
(Gleeson and Greder) 140
'group-then-regroup' strategy 11
groupings, for classroom work 10; for
8–10 years 140–1, 143; for 10–12 years
148–53
Gutenberg Project 3
Guyer, C. and Joyce, M. *Lasting Image*
112

Harry Potter and the Chamber of Secrets
game (Electronic Arts) 125, 133
Harry Potter books: 'fan' sites 43, 53;
J.K. Rowling website 38–9
Harry Potter games 124, 125
Haunted Castle, The 105
Henry P. Baloney (Scieszka and Smith) 42
Holes (Sachar) 155–6
Hot House (Rock) 107, 108–13
hot spots *see* hyperlinks
Hunt, P. 54, 107, 156
hyperlinks (hot spots): in book-based
literary narratives 58, 59; in digital
poetry 114–15; in digitally originated
literary text 4, 5(*Figure 1.1*), 87–90; in
e-poetry 115; in e-stories for early
readers 90; in hypermedia narratives
107–8, 113; in hypertext narratives
103–4; in re-contextualized literary
texts 65–6, 68; in re-contextualized
versions of classic stories 85
hypermedia narratives 4, 88, 89(*Figure
5.1*), 105–14
hyperpoetry 88, 115
hypertext narratives 4, 88, 89(*Figure 5.1*),
102–5

ideational structures *see* representational meanings
images: in CD-Rom versions of classic stories 73–6, 85; in CD-ROM versions of literary texts 3–4, 58–71(*Figures 4.1–4.2*); on 'fan' sites 52; in hypermedia narratives 111–13; in interactive story contexts 99; in linear e-narratives 95–8; and the literary text 5–7; reading narrative images 16–26; on student-produced websites 49–50; for *The Little Prince* 147–8, 150; for *The Selfish Giant* 138–9, 143–4; in work program tasks 149, 150, 152, 153, *see also* grammar of visual design; point of view; verbal text
Imaginarium site 138–9
information value, of page layout 26
Inner Circle, The (online book) 100–1(*Table 5.1*), 102
InnerWorkings: *The Jolly Post Office* 126
integrated puzzles/tasks 127–9
inter-generational literature 148–9
inter-modal relations, in *Vasalisa* 111–13
interactive meaning, in images 15, 19–22, 63–5(*Table 4.2*)
interactive story contexts: and e-narratives 4, 87–8, 89(*Figure 5.1*), 98–105, *see also* electronic games narratives
International Children's Digital Library 72
international literature 148–9
interpersonal interaction, visual grammar for 19
interpersonal meanings, in functional grammar 27, 28, 31–3
Interpersonal Theme 35
interpretation, contexts of 45–51, 85
interpretative response 8–9; in hypermedia narratives 108
invitation 41–3

Jack and the Beanstalk 139, 140, 141
James, R. 106
Jennings, P.: website 41, 42; *Wicked* stories 42
Jewitt, C. 58
'jig-saw' group work 140–1, 143
jig-saw puzzles 132
J.K. Rowling website 38–9
Jolly Post Office, The (InnerWorkings) 126

Jolly Postman, The (Ahlberg and Ahlberg) 126
Joyce, M. and Guyer, C. *Lasting Image* 112
Joyce, W., *George Shrinks* 58–60, 85, 134

Der Kleine Prinz website 149
Knowles, M. and Malmkjaer, K. 31, 35
Kress, G. and van Leeuwen, T. 16, 19, 20, 26, 27

language: foreign language websites 148–9; use of in *Vasalisa* 108–13, *see also* metalanguage; narrative technique; point of view
Larsen, D., *Stained Glass Windows* 115
Lasting Image (Guyer and Joyce) 112
layout *see* compositional meanings
learning activities *see* classroom work
learning programs *see* programs of work
lesson plans, online 45–6
Leu, D. 156
Lewis, K. *see Wolstencroft the Bear*
libraries, online 3
lineality, in hypermedia narratives 107, 108
linear e-narratives 4, 87, 89(*Figure 5.1*), 94–8
literacy development, and electronic games narratives 129–35; *see also* reading
literary texts *see* books (printed); digitally originated literary text
Little Prince, The (de Saint-Exupery) 23–6; CD-ROM and book versions compared 60–5(*Figures 4.1–4.2, Tables 4.1–4.2*), 66(*Table 4.3*), 85, 148; 'fan' site 44; the Fox Taming Game 126–7, 131–2; as group work program 147–53
Littlest Knight, The (Moore) 20, 22(*Figure 2.4*), 23, 24(*Figure 2.5*), 97(*Figure 5.5*); point of view in 95, 150
Locke, T. 156
Lord of the Rings Fantasy World website 130
Lord of the Rings: The Return of the King game (Electronic Arts) 125
Lothian books, website 45
Luke's Way of Looking (Wheatley and Ottley) 149
Lulu's Enchanted Book (Victor-Pujebet) 106, 151–2, 153

McCloud, S. 18, 116
Mackey, M. 1, 38, 133, 156
Marked Themes 36
Marsden, J., *The Rabbits* (Marsden and Tan) 36, 45
Marsden, J. *The Wire* 20–2
Mascarenhas, A. 42
Material Processes (action verbs) 27, 30, 77–8(*Table 4.5*)
Mattell Media, *Arthur's Absolutely Fun Day* game 133–4
Matus, M., *The Relic Triangle Trilogy* 98, 100–2
Megan's Star (Baillie) 151
Mental Processes (thinking/feeling verbs) 27, 29, 91, 92(*Figure 5.2*)
Merino, R. *The Farm Animals* 90
metalanguage 13–16; and a 'grammar of visual design' 16–26; taught in work programs 143–4, 145, 147–8, 150, *see also* narrative techniques; point of view
Middle Earth palace site 52, 53–4
Miller, L. 49
modality 20–2
mood and modality 31–3
Moore, C., *The Littlest Knight* 20, 22(*Figure 2.4*), 23, 24(*Figure 2.5*), 97(*Figure 5.5*); point of view in 95, 97(*Figure 5.5*), 150
Moore, C.: and Paulhamus, J., *Second Thoughts* 96–8; *Wind Song* 20, 21(*Figure 2.2–2.3*), 95, *see also Wolstencroft the Bear*
Mulan, CD-ROM and book versions compared 68–71
Mulan Animated Storybook (CD-ROM) (Disney) 30, 68–9, 70–1, 125, 129
music *see* sound
My Obsession with Chess website (McCloud) 18, 116

narrative images *see* images
narrative techniques: CD-Rom and printed texts compared 69–70, 75–84; in electronic games 121–2, 122–9, 130–1; *see also* textual meanings
naturalistic images 20–2
navigation, in *The Ballad of Sand and Harry Soot* 115
Neale, S. 138
Netlibris site 46–7
New London Group 14, 15

New South Wales Department of Education and Training (NSW DET) 48
New Zealand, Book Chat site 48
Newman, M. 45
Ng, *Tiger Son* 18
Nightmare Room website 98, 100
Nodelman, P. 6–7

obernewtyn.net club 51–2
O'Brien, R., *Z for Zachariah* 40, 49, 134
Of Mice and Men (Steinbeck) 58
'offers' 19
Ollila, M. 46
online games *see* electronic games narratives
online lesson plans 45–6
online reading, for work programs 149
organizational frameworks 2
otherworldly characters 149–50, 150–1
Ottley, M., *Luke's Way of Looking* (Wheatley and Ottley) 149

palace sites 52–4
Palace User software 53
Participants 27, 28
Pavic, M. *Glass Snail* 104
Payuta and the Ice God (Ubisoft) 106
pedagogic frameworks 2; for use of e-literature 7–11
Penguin Readers page 139, 142
'phase space' 38
Philip Pullman website 38, 41
Philosopher's Stone, The (Rowling), versions of 39
photography, used in work programs 145
Piggybook (Browne) 18, 28–9
players, virtual identity of 121–2; *see also* characters
poetry, digital 88, 114–15
point of view: and interpretation of narrative 95–8(*Figures 5.4–5.5*); in *The Little Prince* 19–21(*Figures 2.2–2.3*), 63–5, 66(*Table 4.3*); in *The Littlest Knight* 95, 97(*Figure 5.5*), 150; in *Vasalisa* 111; in *Wolstencroft the Bear* 91–4; in work program tasks 150
Polar Express Reading Challenge 155
Ponzio, R. and Labate, J., *Flying Angels* 94
Ponzio, R. *St Mary's Avenue* 94
Porphyria's Lover website (McCloud) 116
Powerpoint 117
Present Takers, The (Chambers) 39–40

primary education 46
printed book format *see* books (printed)
Processes (verbs) 27–31, 70–1, 77–8(*Table 4.5*), 91, 92(*Figure 5.2*); in *Vasalisa* 110–11
programs of work 9–11, 137–8; online lesson plans 45–6; *The Selfish Giant* 138–46; websites for classroom programs 138–9, *see also* classroom work
publication, online: online re-publication 3; of students' work 49, *see also* electronically re-contextualized literary texts
publishers, websites of 38–41, 55, 155
Pullman, P.: *His Dark Materials* 38; *Lyra's Oxford* 38; website 38, 41
puzzles: adjunct 125–7; integrated 127–8

Queensland English 1–10 Syllabus 31
Queensland University of Technology (QUT) 48

Rabbits, The (Marsden and Tan) 36, 45
Random House, Philip Pullman website 38, 41
reactional processes 17–18
Read-Write-Think site 40
reader alignment 91, 93(*Figure 5.3*), 97(*Figure 5.5*), *see also* point of view
Readers Theatre 139, 146
reading: enticements for 8, 155; and game-playing 133–4; methods of in e-poetry 115; online, in work programs 149, *see also* literacy; students
Red Badge of Courage, The (Crane) 40
Relic Triangle Trilogy, The (Matus) 98, 100–2
representational meanings: in functional grammar 27, 28–31; in images 15, 16–19
reviews, online 43, 46
Rheme 34, 35
Richards, C. 6
Right Number, The website (McCloud) 116
Roald Dahl website 46
Rock, J. 107
Rose Meets Mr Wintergarten (Graham) 140, 142–3
Rowling, J.K.: 'fan' site 43; *The Philosopher's Stone* 39; website 38–9

Rubinstein, Gillian, 'fan' site 44
Russell, P.C. 139
Rutgers State University, site for children's literature 54–5

Sachar, L., *Holes* 155–6
St Mary's Avenue (Ponzio) 94
salience 23, 26
scaffolding learning tasks 11
Scieszka, J. and Smith, L.: *Henry P. Baloney* 42; *The Stinky Cheeseman and Other Fairly Stupid Tales* 42
Scott McCloud site 18, 116–17
Second Thoughts (Moore and Paulhamus) 96–8
Secret Garden, The (Burnett, F.H.) 31, 35–6
Selfish Giant, The (classroom work program) 138–46
Selfish Giant, The (Wilde) 138–9
Sendak, M., *Where the Wild Things Are* 17–18
SFG (systematic functional grammar) 27
SFL (systematic functional linguistics) 7, 15
Sherlock Holmes games 134
Shrek e-book (Ashby) 32–3, 35
Shrek storybook (Steig) 32
sites *see* websites
Skellig (Almond) 39, 47, 149–50, 151, 152, 153
Slawek website 148
Snow White and the Seven Hansels (Tivola) 129
social distance 19–20
software 50, 53, 117
Song of Mulan (Lee) 30, 69
Sorensen, L. 53
sound: in e-stories for early readers 90; in hypertext narratives 103
'speech bubbles' 18
Spywatch (BBC) 47, 123–4
SQUARE-ENIX, *Final Fantasy X* game 134–5
Stacks, The website 105
Stained Glass Windows (Larsen) 115
Steinbeck, J., *Of Mice and Men* 58
Stellaluna (Cannon) 58, 65–8
Steve Conley's Astounding Space Thrills (e-comic) 116
Sticky Burr (e-comic) 116
Stig of the Dump (Topologika) game 125

Stine R.L., The Nightmare Room website 98, 100
Stinky Cheeseman and Other Fairly Stupid Tales (Scieszka and Smith) 42
stories: CD-ROM contextualizations of classic stories 71–85; in CD-ROM versions of literary texts 3–4, 58–71, 151–2; e-narratives 4, 87–8; video games as 'embodied' stories 120
story genesis *see* composition
story-focused games 122, 124–9, 131–3
Storysocks site 138
Street, *The Geode Space* game 127, 132
Strickland, S., *The Ballad of Sand and Harry Soot* 115
students: book raps for 48–9; early readers' e-stories 87; expertise of 11–12; games for different age groups 133–4; student-produced websites 49–50, 138; work programs for different age groups 137–8
Sundog Stories website 94
Swigart, R., *About Time* 26, 102–4
syllabus documents, functional grammar in 36
symbolic images 18

'talking books' (CD-ROMs) 106
Tan S., *The Rabbits* (Marsden and Tan) 36, 45
tasks *see* puzzles
teachers: online discussion management 47, 48; role and expertise of 11–12; website resources for 40, 55
text: book and electronic versions compared 67–8; and images 5–7
textual meanings, and functional grammar 27, 28, 33–6, 82–4
Theatrical Resources Home Page 139, 146
Themes 33–6, 79–84(*Tables 4.6–4.7, Figure 4.6*)
thinking verbs, visual grammars for 18, 27, 91, 92(*Figure 5.2*)
Thomas, A. 47, 53
Tiger Son (Ng) 18
Tivola, *Snow White and the Seven Hansels* game 129
Tolkein stories, 'fan' sites for 53–4
Tomb Raider: The Last Revelation (video game) 121–2
Topical Themes 33–5, 79–84(*Tables 4.6–4.7, Figure 4.6*)

transactional actions 17
Tumblebooks site 90

Ubisoft Multimedia, *Payuta and the Ice God* 106
Uncle David (Gleeson) 140, 141
University of Calgary children's literature site 54
USA, E-pal project 48

Vandergrift's children's literature page 54–5
Vasalisa Electric (Rock) 107, 108
Vasalisa Project, The 107–8(*Figure 5.6*)
Vasalisa story (Rock) 26, 108–13
verbal processes 18, 27, 28–9
verbal texts: in hypertext narratives 102–5; language use in *Vasalisa* 108–13
verbs *see* Processes
Veteran, The (Crane) 40
Victor-Pujebet, V., *Lulu's Enchanted Book* 106, 151–2, 153
video clips: identification with characters 121–2; of novels 45–6
video games 119, 120–2, *see also* electronic games narratives
viewpoint *see* point of view
virtual identity 121–2
visual grammar *see* metalanguage; point of view
visual imagery *see* images
Volotinen, T. 46

War of the Worlds, The (Wells) 40
Warner Brothers website 100, 155
websites: authors' and publishers' 38–41, 46, 55, 98, 100, 155–6; for bloggers 51; for classroom work sessions 138–9, 142–3, 145, 146, 147, 152; for digital poetry 114, 115; for e-narratives 90–1, 94, 98, 99; for electronic games 123, 124, 130, 134; 'fan' sites 4, 5, 43–4; for hypermedia narratives 107–8, 113–14; for hypertext narratives 102, 104, 105; offering learning activities 45–6; for online communication 46–51; specialist sites 54–5; student-produced 49–50, 146; for teachers 40, 55; for *The Little Prince* 147, 148–9; for *The Selfish Giant* 138, 139; for young people 13, 54–5
Wells, H.G., *The War of the Worlds* 40

Wheatley, N., *Luke's Way of Looking* (Wheatley and Ottley) 149

Where the Wild Things Are (Sendak) 17–18

Wicked stories (Jennings and Gleitzman) 42

Wilde, O. *The Selfish Giant* 138–9

William Tell (Early) 17, 18, 26, 30, 32; CD-ROM and book versions compared 72–84(*Figures 4.4–5.5*), 72–84(*Figures 4.4–5.5, Tables 4.5–4.7*)

William Tell Told Again (Wodehouse) 20, 30, 32, 33–5; CD-ROM and book versions compared 72–84(*Figure 4.3, 4.5 and 4.6, Tables 4.4–4.7*)

Wind Song (Moore) 20, 21(*Figure 2.2–2.3*), 95

Wishing Cupboard, The game (Hathorn) 127–8, 132

Wodehouse, P.G., *William Tell Told Again* 20, 30, 32, 33–5; CD-ROM and book versions compared 72–84(*Figures 4.3, 4.5 and 4.6, Tables 4.4–4.7*)

Wolstencroft the Bear (Lewis and Moore): Mental Processes in 29; online versions of 91–4(*Figures 5.2, 5.3* and *5.4*); visual grammar in 16–17(*Figure 2.1*), 19, 20, 23, 25(*Figure 2.6*), 26

Woolman, S. (illustrator), *The Dreamwalker* (Carmody and Woolman) 45, 134, 148, 149, 152, 153

word order 33

Wordcircuits website 102, 104, 113–14, 115

work programs *see* programs of work

Wright, F. 138

Wright, J. *Ant Lion, The* 31

Wrightson, Patricia, 'fan' site' 44

writing (stories) *see* composition

Xylo (Howard) 114

Your Own Personal Nightmare (Stine) 100

Z for Zachariah (O'Brien) 40, 49, 134

Zervos, K., *Childhood in Richmond* 113–14

Zhang, N., *Ballad of Mulan* 41, 69–70

Zoo (Browne) 18

eBooks – at www.eBookstore.tandf.co.uk

A library at your fingertips!

eBooks are electronic versions of printed books. You can store them on your PC/laptop or browse them online.

They have advantages for anyone needing rapid access to a wide variety of published, copyright information.

eBooks can help your research by enabling you to bookmark chapters, annotate text and use instant searches to find specific words or phrases. Several eBook files would fit on even a small laptop or PDA.

NEW: Save money by eSubscribing: cheap, online access to any eBook for as long as you need it.

Annual subscription packages

We now offer special low-cost bulk subscriptions to packages of eBooks in certain subject areas. These are available to libraries or to individuals.

For more information please contact webmaster.ebooks@tandf.co.uk

We're continually developing the eBook concept, so keep up to date by visiting the website.

www.eBookstore.tandf.co.uk